A E Durrant

AUSTRALIAN STEAM

A. H. & A. W. REED
SYDNEY WELLINGTON AUCKLAND

To the memory of John Hogg, Geoffrey Sweet and Russell Robertson, who died tragically in a road accident whilst *en route* to witness the brief steam revival in New South Wales, July 1972.

Originally Published by David & Charles Limited
First Published in Australia and New Zealand 1978
A. H. & A. W. Reed Pty Limited
53 Myoora Road Terry Hills Sydney 2084
65–67 Taranaki Street Wellington
16–18 Beresford Street Auckland
Cnr Mowbray & Thackeray Streets Waltham Christchurch

ISBN 50027 9

Printed and bound in Great Britain

CONTENTS

LOCOMOTIVE STUDIES

Published Titles

The Steam Locomotives of Eastern Europe, by A. E. Durrant
Steam Locomotives in Industry, by The Industrial Locomotives Society
Steam Locomotives of the South African Railways, Vols 1 and 2, by D. F. Holland
Steam Locomotives of the East African Railways, by R. Ramaer
The British Internal-Combustion Locomotive: 1894–1940, by Brian Webb
Diesel-hydraulic Locomotives of the Western Region, by Brian Reed
English Electric Main Line Diesels of BR, by Brian Webb
The Drummond Greyhounds of the LSWR, by D. L. Bradley
The Stanier 4–6–0s of the LMS, by J. W. P. Rowledge and B. Reed
The Garratt Locomotive, by A. E. Durrant
The Mallet Locomotive, by A. E. Durrant
The Fairlie Locomotive, by R. A. S. Abbott
4–8–0 Tender Engines, by D. R. Carling

LOCOMOTIVE MONOGRAPHS

General Editor:
O. S. Nock, BSc, CEng, FICE, FIMechE
The GWR Stars, Castles & Kings, Parts 1 and 2, by O. S. Nock
The LNWR Precursor Family, by O. S. Nock
The Stirling Singles of the Great Northern Railway, by K. H. Leech and M. Boddy
Gresley Pacifics, Parts 1 and 2, by O. S. Nock
The Southern King Arthur Family by O. S. Nock
The GWR Standard Gauge 4–4–0s, Vols 1 and 2, by O. S. Nock
Royal Scots & Patriots of the LMS, by O. S. Nock

FOREWORD

AUSTRALIAN STEAM is the first book published covering the steam locomotives of Australia as a complete entity. Although a glance at the bibliography will reveal that there is no dearth of published material dealing with individual sections of the steam story 'down under', the consistency of this material varies from a substantially complete story in the case of New South Wales, through partial coverage in other states, to almost nothing in one or two instances. However, the gaps are slowly and surely being filled, and one day no doubt there will be available a fairly comprehensive set of literature, locally produced, giving extensive coverage of the subject.

Given the above background, how then can we justify this volume, written by a 'bloody Pommie bastard' with but a couple of years' residence in a country already well stocked with steam enthusiasts? The answer lies in the outlook of many average Australians, whereby the strange denizens from states other than his own are regarded with more caution, if not downright suspicion, than people from countries overseas! Scurrilous nicknames have been coined to describe these species native to the 'other states'; to give a few amusing examples, Queenslanders are *banana benders*, Western Australians *sand gropers*, and South Australians *crow eaters*! It follows logically that few, if any, Australian enthusiasts could be relied upon to give an unbiased comment on the inferior locomotives used 'Interstate'—no New South Welshman worthy of the name would admit to the greater size of a Victorian 'S' class compared with the indigenous '38' class pacific, and likewise, no true Victorian would allow that the '38' was more modern in concept than the 'S'. Thus there is a very real place for a book written by the interested outsider, and as there are some hazy notions, both from within and without the country, as to how Australian steam locomotives compare with contemporary examples elsewhere in the world, suitable comparisons have been made to place them in perspective, not only with each other, but on a worldwide basis. The overall picture which emerges is that the best was comparable with the best in many countries of the world, but unfortunately the most modern classes were built in too small numbers to have much overall effect, whilst they were usually too heavy for universal use. Modern power for general use and for light branches was a distinct rarity, and no up-to-date shunting power was ever built.

The story which emerges is quite fascinating, showing as it does that Australian steam locomotives are more interesting than is generally thought overseas, unless one is a technological snob, for the Australian steamer was very much the rugged and reliable workhorse, suited to outback conditions. Modern developments as applied ran more to cast beds than to compounding, and roller bearings rather than complex feed-water heaters, in which overall features American practice of the twentieth century ousted the British of the nineteenth on most of the state systems.

A. E. DURRANT,
Springs, South Africa.

CHAPTER 1

THE AUSTRALIAN SCENE

Australia is one of the largest and most remote countries in the world, situated between the southern sections of the Indian and Pacific Oceans. Popular imagination tends to group Australia and New Zealand as a close clump of islands impossibly distant from any other continent, but this is not really the case, as the following figures show. To start with, New Zealand is not really close to Australia at all, and the direct sea route from Sydney to Wellington is over 1200 miles (2000km) or something like the distance from London to Budapest. On the other hand, the northernmost point of the Cape York Peninsula, Queensland, is less than one hundred miles (160km) from the nearest coastline of the island of New Guinea, in the Indonesian archipelago.

Having settled its proximity from its nearest neighbours, we can add that Australia covers, in round figures, some three million square miles (eight million square kilometres) area, and is 2500 miles (4000km) from east to west, and 2000 miles (3000km) north to south. Administratively, this vast area is divided into six federated states, a regional territory and a Federal capital territory, yet the total population is only between twelve and thirteen million. Fully one-third of this population is accounted for within the two major cities of Sydney and Melbourne, and most of the remainder live in large towns or cities, Australia having the highest percentage of urban dwellers (over ninety per cent) in the world. The typical Australian is not the outback sheep farmer of popular imagination, but a city-dwelling office or factory worker, more at home in an air-conditioned environment than in the open air.

Nevertheless, much of Australia's wealth comes from the country and outback areas, either in the form of minerals or farm produce, and this has necessitated a large mileage of cheaply constructed country branches, usually laid with the bare minimum of engineering work, and having spindly track and earth ballast. The total route mileage in the whole of Australia is substantially more than that in Great Britain (even before Marples and Beeching began their butchery), and about twice that of South Africa, a country with many features similar to Australia, with which a number of useful comparisons may be made. How the area and population is divided up between the member states of the federation, expressed as percentages of the whole, is shown in the table below:

State	*Population* %	*Area* %
New South Wales	36·5	10·5
Victoria	27·5	3·0
Queensland	15·0	22·5
South Australia	10·0	13·0
Western Australia	7·5	33·0
Tasmania	3·0	1·0
Northern Territory	0·5	17·0
	100·0	100·0

New South Wales and Victoria clearly emerge as the most highly-inhabited states, and this is consolidated when one compares the railway mileage and typical locomotive population, again expressed as a percentage of the whole. These strictly railway figures are not absolutely comparable with the state area and population figures, for the Commonwealth Railways operates in three states, using its own locomotives, and owns the mileage operated by the NSWGR in the Australian Capital Territory, but such considerations make a negligible adjustment to the

Symbol of steam survival. The last regular steam workings on any Australian state railway were Garratt-hauled coal trains from Awaba mine to Wangi Power Station. Here the unsynchronised double exhaust from 6019 reverberates from the gum forest, stilling even the kookaburra's mad laughter.

(*A. E. Durrant*)

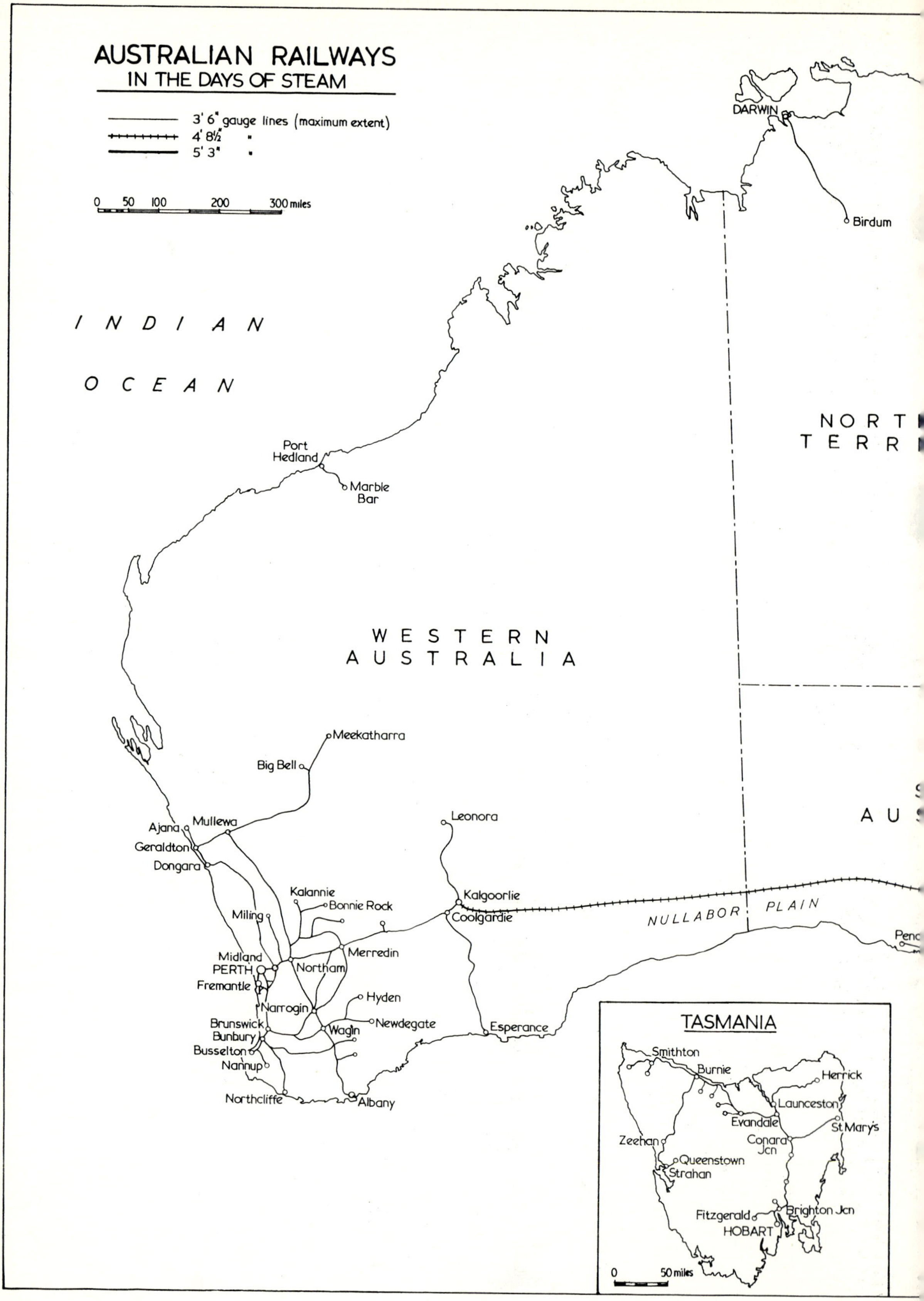

AUSTRALIAN RAILWAYS
IN THE DAYS OF STEAM
3' 6" gauge lines (maximum extent)
4' 8½" "
5' 3" "
0 50 100 200 300 miles
INDIAN OCEAN
DARWIN
Birdum
NORT
TERR
Port Hedland
Marble Bar
WESTERN AUSTRALIA
Meekatharra
Big Bell
Leonora
Mullewa
Ajana
Geraldton
Dongara
Kalannie
Bonnie Rock
Kalgoorlie
Coolgardie
NULLABOR PLAIN
Miling
Midland
PERTH
Fremantle
Merredin
Northam
Hyden
Narrogin
Newdegate
Brunswick
Bunbury
Wagin
Esperance
Busselton
Nannup
Northcliffe
Albany
AUS
TASMANIA
Smithton
Burnie
Herrick
Launceston
Evandale
St Mary's
Zeehan
Conara Jcn
Queenstown
Strahan
Fitzgerald
Brighton Jcn
HOBART
0 50 miles

PACIFIC
OCEAN
ERN
ORY
QUEENSLAND
Rumula
Cairns
Mungana
Normanton
Croydon
Forsayth
Townsville
Merinda
Scottville
Mackay
Rajibbi
Mt Isa
Cloncurry
Hughenden
Dajarra
Winton
Morella
Aramac
Jericho
Blair Atholl
Emerald
Springsure
Rockhampton
Gladstone
Theodore
Yaraka
Bundaberg
Maryborough
Gympie
Quilpie
Charleville
Westgate
Cunnamulla
Injune
Wandoan
Miles
Dalby
Glenmorgan
Toowoomba
BRISBANE
Dirranbandi
Alice Springs
Abminga
OUTH
TRALIA
Maree
Port Augusta
Quorn
Wayalla
Iron Knob
Buckleboo
Port Pirie
Peterborough
Gladstone
Cockburn
Broken Hill
Hope
Moonta
t Lincoln
ADELAIDE
Tailem Bend
Wolseley
Kingston
Mt Gambier
Portland
Mildura
Balranald
Hay
Narrandera
Junee
Cootamundra
Wagga Wagga
Albury
VICTORIA
Bendigo
Ararat
Ballarat
MELBOURNE
Geelong
Woodside
Orbost
Bombala
CANBERRA
Nowra
Wollongong
SYDNEY
Gosford
Bathurst
Orange
Molong
Parkes
Roto
Dubbo
Nyngan
Cobar
Byrock
Bourke
Brewarinna
Pokataroo
Walgett
Moree
NEW SOUTH WALES
Wallangarra
Inverell
Casino
Barraba
Glen Innes
Coffs Harbour
Tamworth
Werris Creek
Maitland
Newcastle

overall picture below:

State Railway	*Mileage* %	*Locomotives* %
New South Wales	23·0	35·0
Victoria	17·0	16·0
Queensland	24·0	24·0
South Australia	10·0	9·0
Western Australia	16·0	11·0
Tasmania	2·0	2·5
Commonwealth	8·0	2·5
	100·0	100·0

From the above, it may be seen that New South Wales emerges as the busiest Australian system, with the highest proportion of locomotives to mileage, Victoria, Queensland, South Australia and Tasmania are par for the Australian course, whilst Western Australia and Commonwealth Railways emerge respectively as slightly and substantially below par. The figures from which these statistics were taken refer to the late 1950s, when steam was still the major item of motive power, but being proportions rather than absolute statistics, a similar broad pattern probably applies in the dieselised 1970s.

We may now have a look at the geographical conditions under which Australian locomotives operate. Australia is possibly the oldest continent on earth, according to some geologists, and its long period of exposure to the elements has resulted in a land which, apart from the narrow coastal strip, is a huge and almost featureless plain, with almost no real mountain ranges. Most of the central regions are dry desert or semi-desert with a hot, dusty, and fly-infested climate, and extensive rainfall is found only along the tropical north coast of Queensland, and Northern Territory, and around the south-eastern coastal districts, especially in Victoria which has a climate calculated to make the immigrant Englishman feel quite at home. Lack of water had a strong effect on steam locomotive operation, and large tenders were generally the rule, with auxiliary tanks or *water gins*, commonly employed in the drier areas and seasons. The Commonwealth Railways, with their routes over the most inhospitable stretches, were particularly sensitive to the water problem, and it comes as no surprise to find them first to dieselise. Coal is available in most states, and comes in variable qualities, with New South Wales and Queensland coal good enough to export, and Victorian coal poor enough to warrant a serious experimental programme on the use of pulverised-fuel firing. Oil firing was only resorted to when strikes made conditions desperate; ironically at the eleventh hour a series of strikes affecting oil supplies caused a partial return to steam traction in New South Wales, although throughout the country dieselisation had progressed too far for there to be any return to steam.

Although much of the country was fairly flat, the policy of cheap railways with minimum earthworks has meant a high proportion of relatively steep gradients, necessitating locomotives with high tractive efforts relative to the weight of train hauled. Furthermore, with track conditions preventing the attainment of high speeds, large wheels were rarely needed, and the archetypal Australian locomotive emerges as a low-wheeled 4-6-0 with outside cylinders, inside valve gear, and narrow firebox, usually very much a 'Jones Goods' in concept. Many of these veterans were later superheated, but it does seem strange that only a few attempts were made to produce a modern lightweight branch locomotive.

The most vexing problem affecting Australia as a whole, and its locomotives in particular, was and is the gauge problem. Until the recent completion of the Sydney–Perth standard gauge transcontinental link, each state went its own way in the choice of rail gauge, each acting in this matter, as with so many others, as though the neighbouring states scarcely existed and were best forgotten. True, when railways were first mooted in Australia, a request was made to Britain as to the best choice of gauge, and the Secretary of State for the Colonies recommended the standard 4ft 8½in used in Britain, Europe and the USA. However, H. W. Shields, chief engineer of the Sydney Railway Company, the first line to get under way, was an Irishman and he persuaded the Australian railway to change to the extravagant Hibernian gauge of 5ft 3in. Under the impression that this was to be the standard gauge of Australia, lines in Victoria, South Australia, and even Tasmania, were planned and started on the broad gauge. Meanwhile, the Sydney people had thought better and reverted to the original proposal of 4ft 8½in, without bothering to inform their southern neighbours, who accordingly continued blissfully building their railways to the 'Irish' gauge. At an early stage, trouble was foreseen, but as nobody was prepared to give way, the tracks of conflicting gauge converged steadily towards the inevitable clash at Albury, making a break of gauge midway between the two most important cities in Australia. When Queensland and Western Australia decided to join in the railway-owning club, 3ft 6in gauge was chosen, and this useful gauge was then adopted by Tasmania which abandoned its broad gauge, by South Australia for a large network of secondary lines, and for the Northern Territory Railway running south from Darwin.

As these railway systems of disparate gauges spread, and eventually met each other, Australia found itself with one of the most muddled collections of unco-ordinated railways that the world has seen. The traveller from Brisbane to Perth, using the most direct route, at one time set off on Queensland's 3ft 6in, changed to standard gauge to cross New South Wales, reverted to 3ft 6in across South Australia to Port Pirie, and then after a brief sampling of standard gauge in South Australia, crossed the Nullarbor Plain on the Trans Australian's standard gauge line before finally ending up on Western Australia's 3ft 6in! Inconvenient enough for the passenger, these breaks of gauge were disastrous when it came to freight traffic, as the transhipment costs made interstate freight prohibitively expensive, isolating the member states of the federation just as surely as thought they were a group of islands separated by the open sea. Australia's small population just could not afford the extravagance of multiple rail gauges and where, for example, a country with uniform gauge and free interchange of freight traffic could produce raw material at A, process it at B, manufacture products at C, and market everywhere from A to Z, Australia had to bear the costs either of heavy transhipment charges for interstate products, or of articles manufactured in uneconomically small quantities for the home state market only. The overall effect of this inefficient national rail system can be seen in the very high cost of living endured by Australian residents, and by the small volume of internal traffic handled by the railways as a whole. Australia may be compared with South Africa in being a southern hemisphere country of similar climate, rich in minerals and with vast areas developed for agriculture by an immigrant, ex-European population. One measure of a country's economic health is the number of heavy and powerful locomotives necessary to prevent hardening of the vital traffic arteries, and as both countries opted for the 4-8-2 and 4-8-4 types, with more than 60 square feet of grate area and more than 47,000lb tractive effort for the heaviest main line haulage, we have but to compare the quantities of these types built for each country. Australia had forty-eight big engines (the New South Wales 57 and 58, South Australian 500, and Victorian 'H' classes), whereas South Africa commissioned no less than 575 comparable locomotives of classes 15E, 15F, 23 and 25! One may inflate the Australian figure by including the Class 60 Garratts, but as this would bring in about four times the quantity of large South African Garratts, each of greater capacity than a NSW 60, the position hardly changes. The inescapable conclusion is that Australia's rail gauge fiasco, fascinating though it is to the railway enthusiast, has, over the years, seriously hampered the economy.

In recent years a small measure of standardisation has been effected such that four of the state capitals have been connected by a transcontinental 4ft 8½in gauge network. This has been effected at considerable expense, and the question arises as to whether 'standard gauge' was the right choice for Australia. To start with, there were more miles of 3ft 6in gauge than any other in Australia, and the lines of this gauge stretched to the four corners of the country. Secondly, it is always less expensive to convert from a broad gauge to a narrower than *vice versa*, and as a smaller mileage is involved in any case, the economical advantages of converting Australia to 3ft 6in seem incontrovertible. Finally, the obvious question looms—can 3ft 6in achieve the results attained by today's limited 4ft 8½in network? Here the answer is in the affirmative, as may be seen from the three sets of comparisons below, detailing respectively Pacific express engines, heavy eight-coupled general-purpose engines, and Garratt articulated locomotives, as supplied to railways in Africa and Australia, showing that whatever capacities were needed in Australia had been met on 3ft 6in gauge in Africa. From the above, it may be deduced that Australia could have handled its heaviest freight and fastest passenger trains with steam locomotives already available on the 3ft 6in gauge, with more ease and less cost than that of trying to make the 4ft 8½in gauge the standard. As it stands now, many of the branch lines will never warrant the cost of gauge widening, and will fall victim to road transport, just as the cream of interstate traffic once fell.

The foregoing is in no way meant to denigrate the fine locomotives which represented the best in Australia, but is merely intended to show that

Pacific express locomotives

Railway	*South African*	*South Australian*	*NSWGR*	*Victorian*
Gauge	3ft 6in	5ft 3in	4ft 8½in	5ft 3in
Class	16E	600	38	S
Cylinders (in)	24 × 28	24 × 28	21·5 × 26	(3) 20·5 × 28
Wheels	6ft 0in	6ft 3in	5ft 9in	6ft 1in
Grate area (sq ft)	62·5	55	47	50
Tractive effort at 85% (lb)	39,980	39,300	36,200	41,670

Heavy 4-8-2 and 4-8-4 locomotives

Railway	*South African*	*South Australian*	*NSWGR*	*South African*	*South Australian*
Gauge	3ft 6in	5ft 3in	4ft 8½in	3ft 6in	5ft 3in
Class	23	500	57	25NC	500B
Type	4-8-2	4-8-2	4-8-2	4-8-4	4-8-4
Cylinders (in)	24 × 28	26 × 28	(3) 23¼ × 28	24 × 28	26 × 28
Wheels	5ft 3in	5ft 3in	5ft 0in	5ft 0in	5ft 3in
Grate area (sq ft)	62·5	66·7	65	70	66·7
Tractive effort at 85% (lb)	48,960	51,000	56,000 (74%)	51,410	51,000 (plus booster 8,000)

Garratt locomotives

Railway	*South African*	*Rhodesian*	*South African*	*East African*	*NSWGR*
Gauge	3ft 6in	3ft 6in	3ft 6in	3ft 3⅜in	4ft 8½in
Class	GMA	20A	GL	59	60
Type (double)	4-8-2	4-8-2	4-8-2	4-8-2	4-8-4
Cylinders (4) (in)	20½ × 26	20 × 26	22 × 26	20½ × 28	19⅞ × 26
Wheels	4ft 6in	4ft 3in	4ft 0in	4ft 6in	4ft 7in
Grate area (sq ft)	63·5	63·1	75	72	63·5
Tractive effort at 85% (lb)	68,800	69,300	89,140	83,350	63,500

with a uniform, preferably 3ft 6in, gauge, Australia's transport constipation would have been eliminated to the overall benefit and prosperity of the country, thus enabling very many more modern and powerful locomotives to be justified and needed than was the case with the fragmented, multi-gauge set-up.

Chapter 2

NEW SOUTH WALES

New South Wales, although not the largest, is the most important state in Australia, having the most industry and the highest population. It is the only state with its main rail system built to the 'standard' gauge of 4ft 8½in, and the only system to have operated more than one thousand locomotives at any time. So far as seniority goes, NSW was beaten by Victoria, whose first railway was opened a year before the Sydney Railway Company commenced operations in 1855. This pioneer NSW line ran from the port of Sydney to Paramatta, then the Governor's country seat, but now an outer suburb of Sydney. A little over a year later, in March 1857, the first section of the Northern system, originally conceived as the Hunter River Railway, was opened from Newcastle to Maitland. For over thirty years the Northern and Southern systems remained physically unconnected, with no rail service between Sydney and the principal provincial city of Newcastle. The obstacle separating the two systems was the Hawkesbury River, bridged in 1889, allowing for the first time an integrated New South Wales Government Railway.

Early locomotives were all of essentially English types. Both the Northern and Southern systems commenced operations with inside-cylinder 0-4-2 tender engines for mixed traffic duties, followed by outside-cylinder 2-2-2s for passenger work. In these early days, engines were classified by the number of the first engine of each type, suffixed by 'N' for those in the Northern division. The first 0-4-2s operating from Sydney were class 1 and the almost identical engines at Newcastle were class 1N. There were a few engines sporting Allan's semi-outside frames of Crewe origin, some 2-2-2 passenger engines used in both Northern and Southern division, 2-4-0s in the south and 4-4-0Ts in the north. None of these were to have any lasting effect on NSW practice and will not be dwelt upon here: anyone desirous of studying these engines further will find illustrations and additional information in the books listed in the bibliography.

It was in 1865 that a very important type was first placed in service for use on the heavy (1 in

Typical of the earliest New South Wales power is 0-4-2 No. 5, built by Hawthorn Leslie in 1855, placed in service 1856, withdrawn during 1890. (*NSWGR*)

30) gradients on the main southern line between Picton and Goulburn, and Robert Stephenson & Co. delivered the first of twenty-three of their patent 'long boiler' 0-6-0, with inside cylinders and all wheels ahead of the firebox. Known as the 17 class, the later examples were locally built, and were followed in 1874–75 by six larger-wheeled examples, class 60, for mountain passenger work. These passenger engines had cylinders with (for those days) the enormous dimen-

Above: Nonagenarian freight engine. 1878 inside-cylinder long-boiler Beyer Peacock 0-6-0 No. 1919 shunts on the shores of Botany Bay 201 years after Captain Cook landed there. (*A. E. Durrant*)

Right: A classic Beyer Peacock 4-4-0T of 1880, the O class was rebuilt as tender locomotives in 1910, ending their days on the Commonwealth Railways. (*NSWGR*)

Below: The New South Wales official 'Vintage Train' on its way to the town centenary at Glen Innes in 1972. Atlas Engineering 4-4-0 No. 176 of 1882 pilots Beyer Peacock 4-6-0 No. 3246 of 1893. (*A. E. Durrant*)

sions of 19in diameter by 28in stroke. Contemporary with the 60 class passenger engines were twelve small-wheeled freight locomotives, substantially larger than the original 17 class, these being class 48. Finally, in 1877 there came the prototype batch of the last and most numerous of New South Wales long boiler 0-6-0s, the 93 class, which eventually totalled some seventy-seven examples. In the 1924 numbering and classification scheme, these became the (Z)19 class, and were retained in slowly-diminishing numbers almost to the very end of steam. After their displacement from main line goods work by heavier power, they found use on lightly-laid branches where they proved powerful, if none too fast. Their short wheelbase, the same feature which limited speeds in branch line service, made them useful for shunting in sharply-curved yards, and a number were employed in Sydney's Darling Harbour until 1970–71. The last two engines of this doughty and long-lived class, 1904 and 1923, were retained at Port Waratah roundhouse, Newcastle, until mid-1972, for pushing wagons up the rustic wooden coal stage, on which no other steam class was allowed.

Contemporary with the long-boiler goods engines came the first of an extended series of four-coupled passenger power, the first of which came out together with the last single-drivers (the 14 class inside cylinder 2-2-2) and the final 0-4-2 series (the 36 class). At the time, the Beyer Peacock 23 class seemed just another variety of passenger engine, but they were one of a great breed of motive power, perhaps the most famous of which were the 4-4-0T used on the Metropolitan and District underground railways in London. This type, with its inclined outside cylinders, is also well known in its narrow gauge form as the Isle of Man 2-4-0T, but in reality there were probably more of this Beyer Peacock classic species in Australia than anywhere else in the world, and we shall come across them frequently in these pages. Reverting to the 23 class 2-4-0s, the first were built in 1865, the year after the 4-4-0Ts were first supplied to the London underground and three years after the type originated on the Tudela & Bilbao Railway, in the north of Spain. The NSW 2-4-0 had a single pair of leading wheels directly under the cylinders. They were displaced from main line passenger services from 1877 onwards when the new 79 class 4-4-0 appeared from Beyer Peacock. The 79 class were a direct development of the Metropolitan tanks, with the same four-wheel leading bissel truck. Sixty-eight locomotives were built to this design, half by Beyer Peacock, and the others by Dübs of Glasgow and Atlas Engineering Works, Sydney. So successful did these prove that the 23 class was subsequently rebuilt with four-wheel bissel trucks to conform, the two classes eventually becoming series (Z)12 and (Z)14 respectively. Like the long boiler 0-6-0s,

Express motive power of the inside-cylinder period. Dübs 4-4-0 No 341 dated from 1885, and became No 1629 before withdrawal in 1929.

(*NSWGR*)

these proved very useful branch line power and about a dozen survived into the 1950s and three into the 1960s. Their longevity was due to good tracking ability over lightly-laid and poorly maintained 'pioneer' branch lines in the outback, where they could safely maintain faster speeds with mail train connections than some of the smaller-wheel branch engines. Two are preserved, 1210 at Canberra station, and 1243 of the locally-built batch, now restored with its original number, 176, kept in working order for the official vintage train, and for enthusiasts' excursions. Surprisingly, when Beyer Peacock were called upon to supply six 4-4-0T class 158 for suburban work in 1880, these were built with inside cylinders, although still with the four-wheel bissel truck. The NSW railways were far from averse to rebuilding engines to suit traffic requirements, and twenty of the 79 class 4-4-0 were rebuilt to 4-4-2T from 1896 to 1902, becoming in the process even more like their Metropolitan ancestors than when built! In Great Britain there were parallels to the NSW engines in both original and rebuilt forms, for the Cambrian Railways in Mid-Wales, having bought four second-hand tanks from the Metropolitan Railway, converted two of them to tender engines, thus providing, on home ground, a pair of NSW 12 class. Even more remarkably, the London & North Western Railway, also the possessor of 'Metropolitan' 4-4-0T engines, converted some to 4-4-2T, producing in the process a NSW 13 class! To complete this chapter of variations, the NSW inside-cylinder 4-4-0T engines were, in 1910, converted to light tender engines for branch line working.

For express work on the more level stretches, where higher speeds were required, the British type inside-cylinder 4-4-0 was adopted and forty-seven of this type supplied from 1883 to 1886. The first six, which eventually became class 15, were by Beyer Peacock, and featured that company's four-wheel bissel truck, as in the earlier outside-cylinder 4-4-0s. The remaining engines came from Dübs; the first twenty-four had the four-wheel bissel truck, and the final seventeen a normal centre-pin bogie, with longer wheelbase. Both the Dübs series became class 16, although they differed quite substantially in the leading bogie. Being less suited to branch line work than the smaller-wheeled, outside-cylinder classes, the 15s and 16s lasted less well; all the Beyer Peacock locomotives were gone by 1931, and the Dübs engines, irrespective of type, by 1933. In their later years they were given Belpaire boilers, and unsightly extended smokeboxes with flat bottoms. Incidentally, the final Dübs engines, those with the longer bogies, had Joy's in place of the original Stephenson valve gear.

The final 4-4-0 type came from the Vulcan Foundry in 1887, and were again of the outside-cylinder type, with smaller wheels, in some ways a development of the earlier Beyer Peacock type. Finally known as the 17 class, they had a long wheelbase, centre-pin bogies, and Stephenson valve gear driving overhead slide valves by means of rocking shafts, American fashion. As built they had extended smokeboxes but later, when reboilered with Belpaire fireboxes, short smokeboxes were fitted. Reputedly, they were heavy on the track, and rough riders, but nevertheless stood the test of time by remaining intact until 1935, when the first was withdrawn, eight out of the twelve surviving until after the second world war. 1709, now restored with its old number, 381, is still in working order for enthusiasts' and vintage train specials.

There is one class of 4-4-0 not yet mentioned, and these were the two American locomotives of class 105. In the late 1870s, enterprising Baldwin salesmen were prowling round the British colonies trying to spread the gospel that American-built engines were more suited to colonial conditions than were British. Strange to relate, the American products of those days, although simple and rugged to the point of crudeness, were small and feeble by British standards, and rarely made much impact upon the assorted colonial authorities. However, reading the correspondence columns of contemporary engineering journals, some of the opposition was blatant prejudice against anything not pure Victorian 'British and best'. In New South Wales, nineteenth-century American power was given a very fair chance to prove its worth, and from 1877 to 1891 some fifty Baldwins of six classes were placed in service. Strangely enough, just as American power at the turn of the century was really beginning to develop some superior features (adopted even in England by Churchward of the Great Western Railway), NSW abandoned Baldwin and relurned to the concept of the classical English tocomotive.

In the meantime, as mentioned, there were fifty Baldwins supplied, commencing with a real 'wild west' 4-4-0, no 105, delivered in 1877. Diamond stack, giant oil headlamp, and boiler-mounted bell all proclaimed the country of origin, although the side buffers and lack of cowcatcher reflected the British standards used in NSW. A similar locomotive was purchased in 1879, plus twelve 2-8-0s, the first eight-coupled power on the line. These were as typically American as the 4-4-0s, and were of the early consolidation type with drive on the second axle. The two 4-4-0s were scrapped in 1904, but the freight engines, later class 28, lasted until 1925–36. The other popular type in America at the time was the 2-6-0, already represented in NSW by a Beyer Peacock design of which more later. In 1885 Baldwin delivered twenty Moguls, ten with small wheels for freight, and ten with larger wheels for passenger service over the Blue Mountains. Most details were similar and typically American, but the passenger engines inexplicably had domeless boilers. Five years later, Dübs delivered ten almost identical locomotives, termed 'Scotch Yankees', for passenger work, these having the running board curved over the driving wheels rather like certain old Great Western broad gauge engines. The goods moguls were scrapped between 1915 and 1927, the 'Scotch Yankees', by then class 22, in 1924–34, and the original Baldwin passenger locomotives mostly lasted until 1938–39. The final two batches of Baldwins came out in 1891, and were a stopgap measure pending the arrival of new engines of similar power from Beyer Peacock, Baldwins offering their usual quick delivery. Twelve were 4-6-0 passenger engines, the first of this soon to be universally popular type in NSW, and twenty were 2-8-0s, larger than the 1885 batch and with the drive on the third coupled axle. All the 4-6-0s came out as saturated slide valve engines, and were later superheated and fitted with piston valves. Of the 2-8-0s, two were built as Vauclain four-cylinder compounds, the rest being similar to the passenger engines in mechanical features. The Vauclains suffered the usual troubles of their type, and were soon converted to simples, and six of the original simples were later superheated and given new piston valve cylinders. One of these was provided with Southern valve gear, prior to the introduction of the 55 class, also with that fitting.

In dealing with the Baldwin engines, we have digressed from the generally chronological order of this narrative, and so return now to 1882 when Beyer Peacock delivered the first of a class which was to have a great impact on NSW power. This was the 205 (later 25) class, a small-wheeled freight 2-6-0. Seventy of these were in service by the end of 1885, replacing the long boiler 0-6-0s on the heaviest duties. Another twenty-five very similar engines, class 24, came out from Dübs, and were placed in service during 1891. The features of these two classes, the low-slung outside cylinders, plate frames, and inside valve gear were to become the standard for freight power until the very end of steam, for developed directly from the Moguls were the 'Standard goods' 2-8-0s designed by W. Thow, as part of the NSWGRs 'great leap forward', just before the turn of the century.

However, before dealing with the Thow regime, it is as well to mention briefly three classes of tank engine built for local work. The first were

The last four-coupled main line locomotives for New South Wales were the Vulcan Foundry H class of 1887. Preserved No 381 is seen here at Waterfall in 1972 with a special bound for Wollongong.
(*A. E. Durrant*)

Another long-lived design. 2-4-0T No 1042, built by Vale in 1887 to a Beyer Peacock design, was works shunter at Cardiff, near Newcastle, until 1973.
(*A. E. Durrant*)

eighteen Beyer Peacock inside-cylinder 2-4-0T, of their classical design supplied to a number of railways, including those on the Isle of Wight. Twelve of these came out from Gorton in 1885, and the other six were built locally, by Henry Vale, in 1887. Due to a derailment at the Sydney suburb of Sydenham, they were taken off passenger work in 1901, although in view of their success elsewhere, it would seem that the unfortunate 2-4-0Ts were selected as convenient scapegoats. Removed from their intended duties so early, these neat little engines have survived with surprising tenacity, 1042, the works shunter at Cardiff locomotive works near Newcastle, lasting until January 1973, almost outliving main line steam. For the steeply-graded North Shore suburban trains, from Milson's Point to Hornsby, Beyer Peacock produced fifteen inside-cylinder 4-4-2Ts in 1891. Lacking in adhesion weight and tractive effort, they were not a great success, and were replaced by larger power as soon as possible, although used on lesser duties and not withdrawn until 1918–27.

There was also a demand for tank engines for hauling short-distance mineral traffic, and two designs were built, each based on existing tender engines, and possessing unusual features. First were twelve Beyer Peacock 2-6-4T, most components being standard with the final (19) class of long boiler 0-6-0. These were produced for the Newcastle coal traffic, and a further twenty-one were later added by rebuilding some of the tender engines into tanks. The unusual feature was the use of inside cylinders on a 2-6-4T. Being very handy, light, and relatively powerful, they survived for a long time as branch line engines, and were particularly noted for their use on the Camden branch, where up to three were needed to haul a train over the 1 in 20 (5 per cent) gradients. Being so useful, they lasted well, the final one being withdrawn early in 1968.

The other class of mineral tank engine, of slightly more capacity, was built by Dübs in 1892, a 2-6-2ST version of the 24 class 2-6-0. In this case, the unusual feature was the use of saddle tanks on a 2-6-2T design. Their principal spheres of operation were on the Illawarra line (Sydney to Wollongong) and in the Lithgow and

Restored 4-6-0 No 3203 at Thirroul in 1972, returning to Sydney with an excursion from Kiama. This was a very advanced design when introduced in 1892.
(*A. E. Durrant*)

Bathurst area. The whole class was still at work in 1956, the last two engines, used for shunting and banking at Bathurst, not withdrawn until 1971. There were also a number of odd small tank engines of no particular interest, with which it is not proposed to deal here.

The Thow period

In 1888 Mr. W. Thow, a former LNWR man, Locomotive Engineer of the South Australian Railways, was called upon to report on the rolling stock position in New South Wales, then considered unsatisfactory. As a result, he was appointed Locomotive Engineer in NSW in 1889, and retitled Chief Mechanical Engineer in 1891. Thow was clearly a 'big engine' man, and he set forth to provide NSW with a set of locomotives, thoroughly British in concept, which were head and shoulders above contemporary British power in size and capacity. Indeed, at their time of introduction, they would have been thought of as big engines anywhere in the world, including the USA. Thow's first NSW design, his 4-6-0 passenger locomotive, was a direct enlargement and development of his 'R' class, previously built for South Australia. Thow must have wasted no time after his appointment in getting these designed and built, and what is more, he had the courage to order fifty as the first batch! Beyer Peacock built all of these, the first at the end of 1891, so that they were shipped to NSW in time to be placed in service in 1892–93. All fifty were in service the year before David Jones' famous goods design, Britain's first 4-6-0, was built. It was remarkably similar in size, features, and appearance. Jones' 4-6-0 was quite different to his previous designs for the Highland Railway, all of which featured the archaic Allan-type framing, and one is tempted to conjecture that Jones, or his chief draughtsman, was sufficiently impressed with the NSW locomotive to attempt something similar. Certainly there was time enough after the appearance of Thow's design for Jones to use it as his inspiration, but we will probably never know the truth of the matter, especially as no Scotsman with a sufficiently intimate knowledge of Highland Railway affairs would ever admit to their most famous design having been inspired by a colonial machine! It is of particular interest to compare the dimensions of the Thow and Jones 4-6-0s, and these are tabulated below:

Railway	*New South Wales*	*Highland*
Date	1891	1894
Cylinders (in)	20 × 26	20 × 26
Driving wheels	5ft 0in	5ft 3in
Boiler pressure (lb/sq in)	160	175
Heating surface (sq ft)	1922	1672
Grate area (sq ft)	27·5	22·6
Tractive effort at 85% (lb)	23,600	24,556
Locomotive weight	56·75 tons	56 tons

From the above, it will be seen that Thow's locomotive had considerably more steaming capacity than the Scottish counterpart, which still weighed nearly as much.

Thow's locomotives had plate frames, outside cylinders, and inside Allan straight-link motion, and were the first NSW machines to be built new with Belpaire boilers. Although Beyer Peacock apparently expressed some diffidence as to the possible success of such heavy engines, they were an immediate success, and rendered obsolete all the other engines used on heavy passenger work. At the request of the railway commissioners, a trial was made with compounding, then attracting much attention, and the last two engines of the first batch were turned out as three-cylinder compounds, using a unique arrangement of cylinders. This probably came about by necessity—in order to alter the original design as little as possible, a two-cylinder compound was the obvious answer, but the loading gauge was too tight to accommodate the 31in diameter low-pressure cylinders required. The ingenious result figured out, whether by Thow's staff or by Beyer Peacock now being unknown, was to retain the single high-pressure cylinder on the left-hand side, as on a two-cylinder compound, but to split the low pressure side into two cylinders and mount them one above the other, driving a common crosshead in the Vauclain manner. A von Borries intercepting valve was employed, and the whole contrivance acted as a two-cylinder compound, with two exhaust beats per revolution. Never a success, they were converted to simples after less than ten years' service, and that was the end of compound expansion in New South Wales.

Apart from the two compounds, Thow's 4-6-0s were a tremendous success and multiplied until by 1911 there were 191 in service. In later years they were superheated and fitted with piston valves, and the whole class remained intact until 1956. Most disappeared in the 1960s; 3246 was the engine entrusted to the last regularly-scheduled steam passenger train in Australia, the Newcastle–Singleton local, which ceased steam operation mid-1971. Two were retained in working order for enthusiasts' specials, including 3203, the first of the class actually placed in service.

Following the success of his P class 4-6-0 Thow went back to Beyer Peacock for a freight version, the result being the T class 2-8-0, the first of which was placed in service during June 1896. This class may be said to have come from two different sources—the overall size, and Allan valve gear etc, were clearly from the P class 4-6-0, while the low-slung horizontal cylinders and small wheels derived from the earlier 2-6-0 type. Both ancestors came from Beyer Peacock, so the breed was pure, if a trifle mechanically incestuous. The T class locomotives soon proved their worth, being multiplied by various firms until by 1916 there were 280 of them at work. The Commonwealth railways also adopted both Thow's P and T classes for the Trans-Australian Railway. Less well known is the fact that a batch of T class built by North British were commandeered by the Royal Engineers ROD during the first world war, and used on the supply lines in France. After the war, they were disposed of to the Nord Belge, a Belgian subsidiary of the French Northern Railway, and after the second world war, despite their small numbers and non-standard details, they were still to be found, for a year or two, as Type 76 of the Belgian State Railways.

The NSW T class were as much ahead of contemporary British freight locomotive practice as were the P class on the passenger side. At the time they were introduced, only two British railways owned eight-coupled tender engines: the Barry Railway in South Wales, and the London & North Western Railway. The latter used small, inside-cylinder jobs with 19½in × 24in cylinders and 20 square feet of grate area in a very thin boiler, and were first built in 1892. The Barry engines only came to Britain by accident, having been built by Sharp Stewart for a Norwegian railway, and bought by the Barry company in 1889 after the Swedish and Norwegian company failed to pay for them. These were more com-

Suburban 4-6-4T No 3013 ending its days sold out of service to a coal mine, seen here at Weston with 4-wheeled hoppers from Hebburn Colliery in 1972. (*A. E. Durrant*)

patible with the later NSW locomotives, but not quite so large, as the following table shows:

Railway	*Barry*	*New South Wales*
Type	0-8-0	2-8-0
Date	1889	1896
Cylinders (in)	20 × 26	21 × 26
Driving wheels	4ft 3in	4ft 3in
Boiler pressure (lb/sq in)	150	160
Grate area (sq ft)	22·6	29·7
Locomotive weight (tons)	48·9	65·8
Adhesion weight (tons)	48·9	59·5
Tractive effort at 85% (lb)	26,000	30,600

The NSW T class was thus substantially larger than the nearest British equivalent, and it was not until 1901 that eight-coupled freight power of comparable capacity began to appear on a number of British lines.

Back home in NSW, the T class became known as the 'standard goods' engine, and was the most numerous single class of locomotive to run in Australia. The last seventy-five were built with superheaters and piston valves, to which many of the earlier engines were converted, but right at the end of steam 5069 and 5112 were working from Port Waratah shed, Newcastle, still unsuperheated, and distinguishable from the superheated version by short smokeboxes. These two engines, plus some superheated ones, were working trains from the Maitland coalfield up to the end of 1972.

The third of Thow's excellent designs, simple and rugged to suit the local conditions, was the S class 4-6-4T, for Sydney suburban services. Built from 1903 to 1917, again with the first batch from Beyer Peacock, these were larger passenger tanks than were generally in use on British lines at the time, where usually 2-4-2Ts, 0-4-4Ts, plus a few 0-6-2Ts, were used. In this case, the NSW engines were not greatly ahead of British practice, and were smaller than Churchward's 2-6-2T 3100 which appeared the same year on the GWR. The most noticeable feature of the NSW engine was its large grate area, needed for the high steam demand on the heavily-graded lines from

In September 1972 a strike cut off diesel oil supplies. Seen here near Lisarow is coal-fired veteran 2-8-0 No 5439 piloting a Garratt on a main line freight from Gosford to Newcastle.

(*A. E. Durrant*)

439

Strathfield and Milson's Point up to Hornsby. Apart from the large Churchward engine, which was far in advance of any suburban tank ever used in Australia, the nearest British equivalent of the S class was the 2-6-2T, also by Beyer Peacock, built for the Mersey Railway in 1887, for similar haulage of suburban passenger trains over heavy gradients. The dimensions of the two classes, each with outside cylinders and inside valve gear, were:

Railway	*Mersey*	*New South Wales*
Type	2-6-2T	4-6-4T
Date	1887	1903
Cylinders (in)	$19\frac{1}{2} \times 26$	$18\frac{1}{2} \times 24$
Coupled wheels	4ft 7in	4ft 7in
Boiler pressure (lb/sq in)	150	160
Grate area (sq ft)	24·5	24·0
Tractive effort at 85% (lb)	23,340	20,250
Locomotive weight (tons)	62·4	72·2

The S class was multiplied until it numbered 145 locomotives, all engaged on the Sydney suburban services. To cope with heavier trains, most at one time had enlarged cylinders, higher boiler pressure and increased adhesion weight, but the electrification of the Sydney suburban services in the late 1920s and early 1930s rendered the class largely redundant. Some stayed on the outer suburban trains until the 1960s, and others worked as Sydney station pilots. The last in regular passenger service were those working Toronto trains from Newcastle, up to about 1970. At the end of 1972, there were still two in service at Enfield shed, Sydney, one as the local potterer and the other as Clyde wagon works shunter, among the very last handful of steam on the NSWGR. Of those redundant by electrification, many were converted to tender engines, which will be dealt with later.

William Thow's final design was an enlargement of his P class 4-6-0, with larger driving wheels for faster running. Only five were built, in 1909–10, just before his retirement, and they were known as class N. Although never multiplied, and reputedly rough riders in later years, they nevertheless set the standard of express engines, with six 5ft 9in coupled wheels, developed continuously to the end of steam in NSW. Later known as the 34 class, they were ultimately superheated, and fitted with piston valves, retaining their inside valve gear, and were withdrawn during 1950–57. Thow may have produced only four designs during his tenancy as CME, but each was thoroughly sound and workmanlike, and destined to become the backbone of the system's motive power, some lasting right to the end of steam, in 1972.

From 1911 to 1932, Mr. E. E. Lucy occupied the CME's chair, and a study of the designs produced under his direction poses questions as to how much active interest he actually had in the locomotives designed under his name. As is now generally known, not all chief mechanical engineers were expert locomotive designers, nor did they need to be. A CME might be a good workshop man, a capable administrator, or be primarily interested in any of the many facets of mechanical engineering under his control. The locomotives brought out under Lucy's regime showed no consistency of design features: valve gear earlier inside, later outside, and of three different types, boilers ranging from Belpaire to round top, with both narrow and wide fireboxes. One suspects that locomotive design did not particularly appeal to Lucy, and that this function was left to the chief draughtsmen, of whom there were clearly more than one during his reign.

At first, caution was exercised, and variations on the theme of Thow's 'standard freight' 2-8-0 began to appear. Seventy-five of these built in 1911–16 were superheated, with larger cylinders, bringing the class total up to 280, the most numerous in the whole of Australia, and probably the most numerous in the southern hemisphere. While these were in course of delivery, a modified version known as class TF was built, although nobody seems to know the significance of the F. These had a boiler coned at the rear ring, and externally were noticeable as having a slightly raised running plate, eliminating the splashers of the earlier variety. Allan valve gear was retained, together with the protruding, low-hung outside cylinders, which always looked as though they needed a strong brassiére! All were superheated from the beginning. From 1912 to 1917, some 190 of this class were turned out, and as if that were not enough freight power a further version, the K class, came out in 1918. The K class was an attempt to modernise the standard freight engine by giving it outside valve gear, and piston valves above the cylinders. Given Walschaerts valve gear, they would have been very neat and useful locomotives, but for some unaccountable reason, all were built with the slumsy Southern radial valve gear. Boilers were tapered and superheated, as in the TF class, and some 120 were built by 1925.

Lucy's first passenger engine was the NN class, a designation which earned them the nickname of 'Nannies'. The NNs were an enlargement of the previous N class, with a similar chassis, larger cylinders, and substantially larger boiler, tapered, with a maximum diameter of 6ft 1in. In their original form, they managed to exude a slightly Great Western air, although very few features were really Swindon in detail. Apparently they were poor performers at first, tending to run short of steam on long gradients, but later acquired a good reputation as fast runners, after sundry improvements had been carried out. Exactly what these improvements were seems lost in the mists of time, but they were possibly concerned with valve events and blastpipe/chimney proportions. Certainly they lasted longer than their predecessors—a few of the thirty-five built were withdrawn in 1959, but the rest lasted on into the 1960s, the last being taken out of service in 1968. Half-way through their lives, they were reframed and given new cabs, plus an unsightly deep valance below the running plate. One preserved in working order and painted in a blue livery may be seen on enthusiasts' excursions from time to time.

Lucy's next design, the 36 class 4-6-0, may be considered a direct development of the N class, and was also the first type to appear with the new decimal numbering and classification, replacing the former letter code. It was in fact the work of one A. J. D. Foster, Assistant CME. The 36 class, although still a two-cylinder 4-6-0 with 5ft 9in wheels, was as different from its forebear as possible. Overhead piston valves driven by outside Walschaerts gear replaced the inside valves and gear of the earlier engine, and the boilers were round-topped instead of Belpaire. An altogether more modern and imposing design was the result, and if the 'Nannies' had a slightly Great Western air, the chimney of the 36 was decidedly Lancashire & Yorkshire. It is interesting here to show two comparative tables, the first giving the development of passenger 4-6-0 power in NSW, the N and NN classes becoming 34 and 35 classes respectively.

Class	*N(34)*	*NN(35)*	*36*
Coupled wheels	5ft 9in	5ft 9in	5ft 9in
Cylinders (in)	$21\frac{1}{2}\times26$	$22\frac{1}{2}\times26$	23×26
Working pressure (lb/sq in)	180	180	180
Maximum boiler diameter (inside)	5ft $0\frac{7}{8}$in	6ft 1in	6ft $5\frac{3}{4}$in
Grate area (sq ft)	26·9	30·5	30·5
Tractive effort at 85% (lb)	26,649	29,000	30,500

From the above, it will be seen that progress lay mainly in steam-using rather than steam-generating capacity, except in that larger diameter boilers, increasing progressively from one class to the next, provided a thermal reservoir very useful on short, sharp gradients. The 36 class suffered from a number of teething troubles, but eventually settled down to being thoroughly reliable and popular passenger locomotives. The driving wheels were 1ft 0in smaller in diameter than that used for express work in Britain, but this was suited to NSW conditions, with plenty of heavy gradients and few lines suitable for really fast running. The 36 class was thus more comparable with British fast freight and mixed-traffic designs, and those most nearly comparable were the Urie and Maunsell S15 classes of the London & South Western (later Southern) Railway, and the Great Western 68XX Grange class. Comparative leading dimensions of the three classes are:

Railway	*NSW*	*Southern*	*GWR*
Class	36	S15	68XX
Two cylinders (in)	23×26	$20\frac{1}{2}\times28$	$18\frac{1}{2}\times30$
Coupled wheel diameter	5ft 9in	5ft 7in	5ft 8in
Working pressure (lb/sq in)	180	200	225
Grate area (sq ft)	30·5	28·0	27·1
Tractive effort at 85% (lb)	30,500	29,860	28,875
Locomotive weight (tons)	86·7	79·3	74·0

From the above, it will be seen that the NSW 36 class was larger than its British contemporaries. However, it should be remembered that Britain possessed even larger 4-6-0s, with large driving wheels and multiple cylinders, of which no equivalents existed in Australia.

The 36 class lasted almost to the end of steam in NSW, and most were in later years fitted with new Belpaire boilers, bearing 200lb/sq in pressure, which increased the tractive effort to 33,890lb. No 3616 was fitted with a Giesl ejector in 1957, with results so good that the new CME, well under the influence of American diesel interests, actually suppressed the publication of

The New South Wales Class 36 locomotives were often called 'pigs', possibly due to their chunky looks. Preserved No 3642 accelerates away from Scarborough after exchanging tablets, in 1972.

(A. E. Durrant)

Pre-war heavy power in New South Wales was the Class 57 3-cylinder 4-8-2. This photograph shows the bulky proportions and the massive 2-to-1 beam of the Gresley conjugated valve gear.

(*NSWGR*)

the test results, and cancelled a repeat order for ejectors.

The final locomotive class built and designed under the superintendency of Lucy, although principally the work of Harold Young, chief draughtsman at the time, and later CME, was again quite unlike anything previously in stock. The 57 class 4-8-2 was very much the American engine, and of the era when Alco produced the massive 4-10-2 and 4-12-2 designs, with three cylinders, for the Southern Pacific and Union Pacific Railroads. The 57s sported cast steel beds with separate cylinders, obtained from the USA, *Delta* trailing trucks, and enormous boilers with wide fireboxes. As in the Alco engines, Walschaerts gear actuated the valves for the two outside cylinders, whilst massively constructed Gresley beams drove the centre valve. That such an engine could have come from the same CME as the relatively frail NN class seems unbelievable, but such was the case, nominally at least. With a maximum axle load of nearly 23 tons, the 57s were restricted in operation to the main line west over the Blue Mountains to Wallerawang, southwest to Junee, and down the Illawarra line as far as Thirroul. On a rated tractive effort basis, the 57s were 67 per cent more powerful than a superheated 50 class, but as they were 94 per cent heavier their power/weight ratio was lower; for the heavy slogging work they were called upon to perform, 139 tons of locomotive weight could have been much more effectively utilised. There was no British equivalent to the 57; in fact right to the end of steam, there was no freight locomotive in Great Britain which came anywhere near the size and capacity of the NSW 57, excluding possibly the solitary LNER Garratt, a special-purpose banking engine whose duties were hardly comparable. To get some idea of how the weight of the 57 might have been better utilised, we nose our way almost instinctively to the German 44 class 2-10-0, built for similar duties, on the same rail gauge, and also with three cylinders. The German locomotive did not have the class 57s enormous boiler, butcertainly steamed well and cannot be considered inadequate in that respect. With a 20-ton axle load, a NSW equivalent of a class 44 would have been less restricted in operation than the native class 57, while the whole locomotive would have had less total weight, more adhesion, and a higher effective tractive effort. To forestall any possible objections on running ten-coupled locomotives around sharp curves, we can only say that the standard German 2-10-0 has well proved itself over tightly-curved routes in Turkey and the Balkans, and would never have any difficulties in New South Wales. It remains only to tabulate the relevant principal dimensions and ratios of the two heavy-duty freight types.

Railway	*New South Wales*	*German State*
Class	57	44 (first series)
Type	4-8-2	2-10-0
Three cylinders (in)	$23\frac{1}{4} \times 28$	$23\frac{5}{8} \times 26$
Driving wheels	5ft 0in	4ft 7in
Working pressure (lb/sq in)	200	200
Grate area (sq ft)	65	51
Locomotive weight (tons)	138·7	112
Adhesion weight (tons)	90	98
Tractive effort (lb)	56,000 (at 74%) (limited cut-off) 64,000 (at 85%)	67,700 (at 85%)
Maximum axle load (tons)	22·8	19·8

A 2-10-0 with the same axle load as the 57 class could have been rated at 25 per cent more tractive effort, ie 70,000lb at the 57s limited cut-off, or 80,000lb at 85 per cent, which would have meant a much more effective use of the engine's weight on steep gradients. The 57s were probably most useful as express freight engines on more level sections, and one attempt was made to use the class on the Melbourne express, in order to eliminate double-heading over the hills, but the working was never made regular. Due to their heavy axle load, preventing re-allocation to secondary and branch line service, the 57s were early victims of dieselisation, being replaced by flatulent boxes inferior in both horsepower and tractive effort, which needed operating in multiple to equal the output of a single 57. The big 4-8-2s were withdrawn between 1957 and 1961, and 5711 has been saved for preservation, unfortunately not in working order.

One interesting rebuild came out under Lucy's superintendency, and that was the conversion of sixty-seven members of the 30 class 4-6-4T into 4-6-0 tender engines, following their being rendered redundant by the electrification of Sydney's suburban service. As tender engines, they became remarkably similar to Thow's original R class on the South Australian Railways. Many of the Thow engines were superheated, including some of the rebuilt tanks,

The first five Class 38 pacifics were streamlined. Here No 3801, which ran from Sydney to Perth and back in 1970, tops Hawkmount with an excursion from Newcastle in 1972. (*A. E. Durrant*)

Light 4-6-0 No 3144 making a final run with an enthusiasts' special before being taken out of service in 1972. Built as a suburban 4-6-4T it is seen heading towards Tumbarumba from Wagga Wagga.
(*A. E. Durrant*)

although none of the latter themselves were ever superheated. The rebuilds from 30 class tanks were classified 30T.

After Lucy's retirement in 1933, Mr H. Young became the CME and was the last motive power chief to order steam locomotives. Some ten years elapsed from his appointment to the appearance of his first new design, much of this delay being attributable to the depression. Like so many CMEs, his first product was in the shape of a new express passenger design, and this came out as a very modern and thoroughly American Pacific, class 38. With the exception of the Commonwealth Railways, who never owned any Pacifics, the NSWGR was rather surprisingly the last Australian railway to adopt the 4-6-2 type. However, as far back as 1923, Lucy had proposed the use of such a machine, in a bar-framed design with 6ft 3in driving wheels, again American in concept, and not dissimilar to the South Australian 600 class, already by then on order. However, it was to be twenty years before the type actually appeared in NSW and the eventual 38 class was a far better concept, as one would expect. The 38 had a cast steel bed frame, with integral cylinders, Boxpok driving wheels and roller bearing axleboxes. The first five came out from the Clyde Engineering Company, and had streamlined casings, but the remaining twenty-five, built in the railway's shops at Cardiff and Eveleigh, were unstreamlined. Wheel diameter remained at the previous 5ft 9in, rather than the increased height of the earlier proposal, this being more suited to the heavy gradients in NSW, where 1 in 40 is not uncommon on the main lines. The 38s were superb machines, and tailor-made to suit the short trains and heavy grades prevailing. With no equivalent in British practice, they may perhaps be thought of as a beefed-up, smaller-wheeled Britannia. There is a common belief in NSW that the 38s were larger and more powerful than any British Pacific, but this was certainly not the case, as the following comparative figures reveal:

Railway	*NSW*	*LMS*	*LNER*	*LNER*	*Southern*
Class	38	6220	A1	A2	MN
Wheels	5ft 9in	6ft 9in	6ft 8in	6ft 2in	6ft 2in
Grate area (sq ft)	47	50	50	50	48·5
Pressure (lb/sq in)	245	250	250	250	280
Cylinders, number	2	4	3	3	3
Cylinder size (in)	$21\frac{1}{2} \times 26$	$16\frac{1}{2} \times 28$	19×26	19×26	18×24
Tractive effort at 85% (lb)	36,200	40,000	37,400	40,400	37,500
Locomotive weight (tons)	112·4	108·2	104·1	101·0	97·9

From the above, it can be seen that the only dimension which the 38 excels in is weight, and this is largely attributable to the cast steel bed. It would seem that the 38s were not so economical as the British designs, for despite a tender coal capacity of 14 tons, they were re-coaled, without being detached from the trains, at Demondrille, 241 miles from Sydney, whereas the LNER Pacifics with nine-ton tenders were able to run non-stop from King's Cross to Edinburgh, almost 400 miles. The NSW road was harder going than the LNER, and the 38s were probably designed on the American principles of performance first and economy second. There was certainly no questioning the 38s performance, and in June 1966 the ARHS organised with the NSW railways a comparative test between Pacific 3830 and diesel 42101 over the $36\frac{1}{2}$-mile stretch from Picton to Moss Vale, on the southern line, most of which is on a steady climb of 1 in 75 (1·33 per cent). The load hauled in each case was 300 tons, and the steam locomotive had the apparent disadvantage of lugging an extra 80-odd tons of tender over the grade. The results are published in ARHS bulletin No 348, October 1966, and from these it can be learned that the steam engine covered the trial distance in 47min 10sec at an average speed of 46·5mph, while the diesel took 51min 18sec, averaging only 42·5mph. At two check sections, each of two miles duration, the steam was producing a mean drawbar horsepower of 2,145, compared with 1,400 for the diesel. In summing-up the results, it is rather lamely pointed out that the fuel costs of the steam worked out at approximately $20 compared with $15 for the diesel. What nobody seems to have realised is the diesel only burnt less fuel due to its inferior performance—the steam actually produced 107 horsepower per dollar of fuel, and the diesel only 93. A diesel big enough to equal the steam would, pro rata, cost $23 to run the section. Thus, far too late in the day, figures are available to show that the poor railway wasted its money on short life, high cost, diesel

Finest express power in New South Wales, and probably the finest in Australia, Class 38 pacific No 3813 accelerates from Bowral in 1972 with an excursion to Moss Vale. *(A. E. Durrant)*

power which did not even live up to its much publicised claim of saving fuel costs, even when the very latest diesel was tested against a seventeen-year-old 4-6-2 which never was particularly economical compared with the best in Britain and the Continent of Europe! The last regular express passenger services run by the 38s were the Newcastle expresses from Sydney, up to the end of 1970, the last steam-hauled express trains in Australia.

Young's next design, or rather redesign, was the 58 class heavy 4-8-2, built as a direct development of the 57 class. They were designed without the limited cut-off valve setting of their predecessors, and so the cylinder diameter was accordingly reduced to give the same starting tractive effort, although they had to be worked at a longer cut-off than a 57 to give the same power output at any particular speed. Visually, they were distinguishable by the deep skirting below the running plate, and this helped to conceal their achilles heel, the conjugated drive for the middle piston valve. The 57 class, as already described, used Gresley beams for this purpose and, as so many railways discovered, these were not fully satisfactory. Instead, Young used a modified design in which the small one-to-one lever was retained, and the main Gresley two-to-one lever was replaced by a rack-and-pinion arrangement. Apparently the cure was worse than the disease, and the engines were never considered as good as the 57s. Twenty-five were ordered, but only thirteen actually built, although most of the components had been manufactured. In the ARHS publication, *A Century Plus of Locomotives*, the reason given for discontinuance of manufacture is 'lack of funds', a statement which holds no water. At the same time that the

Modernised version of the Class 57, the Class 58 had rack-and-pinion conjugated motion, and were less satisfactory that the Class 57. The valve gear deficiency was used as an excuse to scrap those built and to discontinue those under construction. (*NSWGR*)

railway had 'no funds' to assemble twelve powerful steam locomotives from already manufactured components, they were able to find the money for far more expensive diesels, of substantially inferior performance. The quoted cost of the thirteen assembled 58 class was $150,000 each, so that allowing a third of this cost for erecting the engines, we come to the very conservative figure of $1,200,000 of public money thrown down the drain due to one man's infatuation with the great diesel confidence trick. Even the unsatisfactory centre valve gear arrangement was no excuse for discontinuing the 58s, which could easily and inexpensively have had Gresley gear or a third set of Walschaerts motion. As it was, the 58s were placed in service during 1950–52 and withdrawn in 1956–57.

Soon after the cessation of the second world war, a shortage of freight power arose in NSW, and rather than purchase some of the large numbers of wartime freight locomotives available in Europe, the NSW railways ordered twenty of the USA standard 2-8-2, as supplied in large numbers to Turkey, India, and other countries, for what should have been an off-the-shelf delivery. However, the railways decided to alter the order to include shorter tenders, to fit onto 60ft turntables. For some reason, this delayed delivery for about six or seven years, the engines not arriving until 1952–53. Exactly why this request should have caused such an inordinate delay seems unknown, but possibly the change in specifications was made at a late date, after the original tenders had been built, resulting in claims from the builders. A further change from the wartime engines was in the provision of a Delta trailing truck, but otherwise they differed only in minor details. All were oil-burners as delivered, at a period when NSW miners were indulging in a strike, but after the oil suppliers, apparently thinking the railways were in their power, increased the price of oil by an unacceptable amount, most were converted to coal firing. The 59 class was a very useful engine, suitable for passenger or freight traffic, although the NSWGR, having little faith in its trackwork, restricted all engines with leading pony trucks to 45mph, thus limiting the use of the 59s on passenger trains. Most of the 59s saw little more than a decade of service as oil-burners, but put in the best part of twenty years' work, the last being withdrawn from service mid-1972, just before the end of steam. To the 59 class belongs the distinction of hauling the last regularly-scheduled passenger trains in Australia which were steam powered. At one time, the Singleton–Newcastle train was thought to have been the last, and when this

After a period when no regular steam passenger trains ran in Australia, mid-1972 often saw the Gosford–Wyong–Gosford local 'squirt' steam-hauled, using a Class 59 off the pickup freight. Here No 5917 is seen heading south near Ourimbah. (*A. E. Durrant*)

went diesel in 1971 it was officially the last steam passenger train in Australia. However, in mid-1972 trains 25X and 188X, between Gosford and Wyong (known locally as the 'Wyong squirt') were often 59 hauled, by the engine otherwise idle between the morning and afternoon pickup freights.

In the same unfortunate category as the 58 class came the NSW Garratts, ordered by Young but delivered during Armstrong's virulent anti-steam regime. The 60 class 4-8-4+4-8-4 were the largest and most powerful locomotives used in Australia, and a total of fifty were ordered. The thought of steam locomotives far more powerful than any of the competing diesels, and light enough to tread the rails on branch lines, apparently spread panic among the diesel interests who, grasping at the tales of teething troubles, largely nothing more than crew unfamiliarity, persuaded the doubtless unreluctant Armstrong to cancel as many as possible. Beyer Peacock, the builders, naturally resisted this, and a compromise was eventually reached whereby three were actually cancelled and five delivered in pieces as spare parts, leaving forty-two Garratts actually placed in service. Cast-steel bed frames, with integral cylinders, roller bearing axleboxes and driving crankpins made the 60 class thoroughly up-to-date and modern Garratts, and after the initial unfamiliarity had worn off they became one of the most useful and versatile locomotives in NSW. One early complaint was that of having to drive 'backwards' with the engine proceeding cab-first; at trade union insistence, many had dual controls fitted so that the driver could face forward when the engine was running backwards. Needless to say, drivers generally sat the normal way round, even when the engine *was* fitted with dual controls! Dual-control engines were recognisable by the initials DC on the buffer beams. Another crew complaint was that when working flat out, the Garratts made so much noise that it was impossible to hear detonators placed on the rails, so sound-conveying tubes rather like old-fashioned speaking tubes were fitted, leading from the outer bogie wheels into the cab. To increase their range of operation, extended bunkers were also fitted to many of the class. The major alteration made to the 60 class was to increase the cylinder diameter to give greater tractive effort, and adjust weight distribution so that more was available for adhesion. In this author's book, *The Garratt Locomotive*, it was stated that only one locomotive was so modified, and although this information came from the NSWGR, it was well out of date when sent. In fact, twenty-nine 60 class had their

The 'big show' in New South Wales was a pair of Garratts storming the 1 in 40 ($2\frac{1}{2}$%) of Fassifern Bank from a standing start. Here two Class 60s hustle coal from Newstan Colliery to the docks at Port Waratah in November 1971. (*A. E. Durrant*)

cylinders enlarged, at various dates between 1957 and 1961, and these were distinguished with + + painted on the buffer beam. The mechanical department never issued new diagrams covering these various modifications, which increased the total weight of the engines from the 254·8 tons quoted by Beyer Peacock to over 264 tons. A comparison between the 60 class and other Garratts has already been made, on page 12, and need not be repeated here.

The NSW Garratts were used fairly widely through the system, on the Western lines in the Molong and Dubbo area, south from Sydney, and on the northern main lines, particularly from Newcastle. In the very last years of steam, the surviving ten were based at Broadmeadow shed, Newcastle, and worked freights down the 'short north' main line to Gosford, coal trains from Newdell (west of Newcastle) to Port Waratah Yards, and more coal from the state mine at Awaba to Wangi power station. Some really spectacular work was performed on these lines, and main line trains from Gosford were often piloted by a 59 class or another Garratt to cope with the 1 in 44 (2·27 per cent) gradient from Dora Creek to Hawkmount, and the shorter stretch of 1 in 40 (2·5 per cent) from Fassifern station. To camp out overnight on the wooded slopes of Hawkmount, and hear the fascinating, syncopated rhythm of a 60s exhaust, as the two units failed to synchronise, was joy to the steam lover's ear. Even more spectacular were the coal trains from Newstan colliery to Port Waratah. Newstan mine was on a short branch from Fassifern, on the 'short north' main line, and normal operation was for one Garratt to bring in empty hoppers and back them into the mine for loading. When the train was nearly full, a second Garratt would arrive light engine, and the two would then double-head the train. The first movement consisted of backing the loaded hoppers out onto the main line at Fassifern station, right at the foot of the 1 in 40 gradient. As the semaphore signal rose to 'clear', two Garratts, billowing smoke from their chimneys and blowing full

A Class 59 2-8-2 teams up with a Garratt to heave a northbound freight over the summit of the $\frac{3}{4}$-mile, mostly of 1-in-40, of Fassifern Bank, in 1972.
(*A. .E. Durrant*)

heads of steam from their safety valves would charge out along the short stretch of nearly level track to gain maximum speed at the foot of the grade. Once on the hill, the pair of articulateds reared up on the 1 in 40 and slower, ever slower, thrashed their way to the summit, situated in a curved cutting. Trees lining the track on each side formed an acoustic trough for sound to blast up the hill ahead of the train, and from the top the decelerating tempo of two 60 class was awe-inspiring. Sometimes, for a few revolutions, the two engines each synchronised in crashing, four-beat explosions, soon to disintigrate into a confused jumble as both units of each locomotive fell out of step.

Lasting as they did to the bitter end, the Garratts participated in two brief 'back to steam' revivals, the first when a strike in the oil industry caused the diesel fuel supply to dry up. For one glorious weekend, the line was again alive with the sight and sound of steam, as double Garratts lifted coal from Newstan, and 59s piloted Garratts on the main line. There were even a couple of night trains double-headed with the old standard goods 2-8-0s. Unfortunately the weekend was marred by the tragic death of three popular and well-known enthusiasts, killed by a speed maniac while driving up to the theatre of steam action.

Soon after, in September 1972, the weekend before the author sailed from Australia, another strike drained Broadmeadow of diesel oil, and as by this time most of the 59 class had been withdrawn, Garratts teamed up with veteran 2-8-0s in what proved to be Australia's last big steam show. The dwindling stocks of 2-8-0s lasted until December, when the miners' holidays threw the locomotives out of work for ever, but a couple of Garratts lingered on hauling coal on the Wangi branch, which was thus the last regular steam operation on an Australian state railway system.

The last Garratt in service was 6042, withdrawn in March 1973. Not only was it the last state locomotive to run, but also the last to be built, by a piece of mechanical and accounting sleight-of-hand. 6042 was the last Garratt assembled anywhere in the world, for in 1969 the original 6042 was worn out, and rather than repair it a new 6042 was built up from the components of 6043–47, delivered in pieces as spares. The new 6042 was sent up to Broadmeadow before the original went away for scrap, and a photograph exists showing the two 6042s side by side. Incidently, in *A Century Plus of Locomotives*, it is stated that the 60 class were the last new steam class introduced on any Australian state railway system. This was not so, for both the South Australian 400 and WAGR V classes were introduced subsequent to the 60 class.

NEW SOUTH WALES: PRINCIPAL DIMENSIONS OF SELECTED CLASSES

1924 Class	*Type*	*Cylinders in×in*	*Wheels ft in*	*Heating Surface sq ft*	*Super-heating Surface sq ft*	*Grate Area sq ft*	*Weight tons*	*Pressure lb/sq in*	*Tractive Effort at at 85% (lb)*	*Old Classes (1855)*	*(1889)*
10	2-4-0T	15 × 22	5 1	900	—	14·0	38·6	140	9,600	351	F
11	4-4-2T	17 × 26	5 1	1276	—	18·5	56·8	160	16,750	—	M
12	4-4-0	18 × 24	5 7	1121	—	14·8	40·7	140	13,800	79	C
13	4-4-2T	18 × 24	5 7	1074	—	14·8	51·9	140	13,800	—	CC
14	4-4-0	18 × 24	5 10	1121	—	14·8	40·4	140	13,200	—	CG
15	4-4-0	17 × 26	6 1½	1123	—	16·9	40·2	150	13,000	255	D
16	4-4-0	18 × 26	6 1½	1123	—	16·9	41·0	140	13,650	261	D
17	4-4-0	18 × 26	5 7	1287	—	20·0	42·8	140	14,950	371	H
18	0-6-0T	15 × 22	4 0	880	—	13·1	37·2	140	12,250	285	R
19	0-6-0	18 × 24	4 0	1294	—	17·9	36·3	150	20,650	93	A
20	2-6-4T	18 × 24	4 0	1322	—	17·9	61·5	150	20,650	—	E
21	2-6-0	18 × 26	5 1	1302	—	17·0	42·5	140	16,400	304	L
22	2-6-0	18 × 26	5 1	1297	—	16·9	47·5	140	16,400	436	L
23	4-6-0	21 × 24	5 1	1594	389	27·7	64·8	160	23,600	—	O
24	2-6-0	18 × 26	4 0½	1408	—	21·0	46·3	140	20,650	—	B
25	2-6-0	18½ × 26	4 0½	941	227	21·0	47·4	140	21,800	205	B
26	2-6-2ST	18 × 26	4 0½	1345	—	21·0	65·9	150	22,150	—	I
27	2-6-0	18 × 24	4 0½	1242	—	21·0	45·1	150	20,450	—	G
28	2-8-0	20 × 24	4 0½	1474	—	28·8	53·4	130	21,800	131	J
29	2-8-0	22 × 26	4 3	1561	402	32·0	63·0	140	29,400	—	—
30	4-6-4T	18½ × 24	4 7	1453	—	24·0	72·2	160	20,350	—	S
30T	4-6-0	19 × 24	4 7	1112	278	24·0	55·2	160	21,420	—	—
32	4-6-0	20 × 26	5 0	1916	—	27·0	56·5	160	23,550	—	P
32	4-6-0	20 × 26	5 0	1485	429	27·0	61·3	160	23,550	—	P
34	4-6-0	21½ × 26	5 9	1623	437	26·9	67·4	180	26,650	—	N
35	4-6-0	22½ × 26	5 9	2235	547	30·5	81·4	180	29,000	—	NN
36	4-6-0	23 × 26	5 9	1990	650	30·5	86·7	180	30,500	—	—
38	4-6-2	21½ × 26	5 9	2614	755	47·0	112·2	245	36,200	—	—
50	2-8-0	21 × 26	4 3	2198	—	29·7	65·8	160	30,600	—	T
53	2-8-0	22 × 26	4 3	1753	364	28·7	71·4	160	33,600	—	TF
55	2-8-0	22 × 26	4 3	1753	364	28·7	73·4	160	33,600	—	K
57	4-8-2	(3) 23¼ × 28	5 0	3390	773	65·0	138·7	200	56,000*	—	—
58	4-8-2	(3) 21½ × 28	5 0	3390	773	65·0	138·7	200	55,000	—	—
59	2-8-2	21 × 28	5 0	2148	624	47·0	89·8	200	35,000	—	—
60	4-8-4 +4-8-4	(4) 19¼ × 26	4 7	3030	750	63·5	255·0	200	59,650	—	—
60++	4-8-4 +4-8-4	(4) 19⅞ × 26	4 7	3030	750	63·5	264·0	200	63,600	—	—

*Tractive effort at 74% boiler pressure.

CHAPTER 3

VICTORIA

The first railway to operate in Australia was the Melbourne & Hobson's Bay line, connecting what was then the most important Australian city with the mouth of the Yarra River, where it flows into the much larger Port Phillip Bay. Today this route still exists as the suburban branch to Port Melbourne, but few of the passengers on the electric trains realise that they are travelling on Australia's oldest railway line. The first engine used was a 2-2-2 well tank, with inside cylinders and plate frames, built locally by the firm of Robertson, Martin, Smith & Co. Until its absorption by the Victorian Railways in 1878, the Hobson's Bay company relied on two main classes of four-coupled tank engine, ten 2-4-0WT built in 1858–70 and six 4-4-0T developments built in 1870–77, all by Robert Stephenson. Each class had outside cylinders and rather vestigial cabs consisting of plates bent over to form roof, front and rear weatherboards. Further examples of the later type were built locally, and were presumably used on the Hobson's Bay line. On the VR the two classes were later known as classes N and C respectively.

The first locomotives of the Victorian Railways came out from George England & Co in 1857, and comprised a 2-2-2 passenger and four 0-6-0 freight locomotives, all with inside cylinders, plate frames, and domeless boilers. The early stock of the VR was unexciting, being in the most part rather feeble versions of the more common types at use in Britain, although a local fitting often applied was the diamond type spark arresting chimney, rarely seen in Great Britain. The passenger engine was number 1, and the 0-6-0s 2 to 4, but soon after their introduction the railway introduced a peculiar numbering system whereby passenger engines carried even, and freight engines odd, numbers.

Beyer Peacock produced the next two batches of engines, five outside-frame, inside-cylinder 2-2-2, and a like number of 0-6-0 freight engines, these latter with inside frames, and similar in appearance to some of McConnell's early designs on the LNWR. George England, and Slaughter, Gruning, built ten 2-4-0ST in 1860, for suburban work, and in the same year the Geelong & Melbourne Railway was taken over, together with its half-dozen outside-cylinder 2-2-2WT of Stephenson design and assorted build. For passenger work, the 2-4-0 soon ousted the single-drivers, and Beyer Peacock built an outside-framed batch of their classic design from 1862, not dissimilar to their famous Dutch engines, plus a smaller inside-framed type for light lines and branches. There was also a 2-4-0 built by the VR at their Williamstown works.

Assorted 0-6-0 freight designs of no real interest came out from various British firms, gradually increasing in size and capacity and, as will be seen later, these were the standard freight type right up to the end of the century, as on most British lines. All had inside cylinders, and most had inside frames as well. In the 1870s, a need was felt for very light engines to operate over the spindly tracks of the lightest country branches which, as a glance at the Victorian railway map will show, consisted of numerous closely-spaced parallel routes. The laying of these lines into the wheat-growing lands was based on limited range of animal-powered transport to and from the farms. It was reckoned that ten miles out and ten back was as much as the farmer could expect from his animals over the rough tracks in use, so the railway lines were built twenty miles apart. Most of these were pioneer lines, built as cheaply as possible to open up the country, and their traffic was largely seasonal, and probably minimal at other times. The engines at first built to run such traffic were curious little outside-cylinder jobs with very small wheels, in fact the type of locomotive normally found on the narrower gauges. All were built locally, either at the Phoenix Foundry, Ballarat, or in the railway's workshops, and they comprised two 0-6-0 classes, later class U, a 2-4-0 (class K) and 4-4-0 (class G). All had inside frames, and slide valves over the cylinders, American fashion, driven from inside gear through rocker arms. The 4-4-0s were the first of their type in Victoria, but soon led to the general adoption of the type for passenger work.

Suburban relic. Class E 2-4-2T No 236 of Victorian Railways would have looked perfectly at home in Victorian England. Here it is seen at North Williamstown Museum. (*A. E. Durrant*)

Another survivor from the Victorian Railways Victorian era is Class Y 0-6-0 No 108, also at North Williamstown. (*A. E. Durrant*)

It was in 1877 that Phoenix turned out eight beautiful little 4-4-0s whose design could only have come from Beyer Peacock. Outside cylinders with overhead valves were the only non-Gorton feature in a class which even copied Beyer Peacock's brass maker's plate round the edge of the driving wheel splasher, and that same firm's elegant copper-capped tapered chimney. In the same year that these were built, Rogers supplied two typical American-pattern 4-4-0s for comparison; neither type was multiplied, and subsequent 4-4-0s were all of the standard British inside-cylinder species.

For Melbourne's growing suburban traffic, Beyer Peacock built an inside-cylinder 4-4-0T, which might be described as an inside connected version of their Metropolitan tank, or perhaps more accurately, as an Isle of Wight 2-4-0T modified to include the firm's short wheelbase, four-wheel bissel truck. Apparently successful, local builders supplied twenty-one more in 1884–86, thus giving the original 1878 engine plenty of time to prove itself. All were rebuilt to 4-4-2Ts in 1900–07, but soon afterwards were replaced on the main services by the later and more powerful 4-6-2Ts.

Meanwhile, American practice for freight power had been considered, and in 1879–83, Baldwin outshopped twelve typical USA-type small-wheeled 4-6-0s. These had long thin boilers with small grates, and could not have been too successful. In fact, American practice of the period was mainly concerned with keeping the engines on the rough track and off the ground, hence the use of a leading bogie on a small-wheeled freight locomotive. Fireboxes were of lesser width and capacity than their British plate-framed counterparts, having to fit between the thicker bar frames, and a smaller proportion of weight was available for adhesion. Phoenix built ten similar engines with larger driving wheels in 1883, and the best use of these engines, later classified S and W respectively, was on the lightly-laid wheat lines. For main lines the British-type inside-cylinder engines continued to be built, and it was nearly twenty years later that a 4-6-0 was built for main line service.

For express and passenger work, the inside-cylinder 4-4-0 persisted, and ten very handsome examples with Beyer Peacock's four-wheel leading bissel came out in 1884, known as the A class. Slower passenger trains were more the rule than expresses in Victoria, hence the twenty D class, built in Ballarat by the Phoenix foundry in 1887–88. These had large, side-window cabs, copied from the American engines, and leading bogies of the more usual centre-pin type, with longer wheelbase. More unusual, the bogies had outside frames, to increase stability, and compensated spring gear. The D class features were then built into two express passenger classes, the 'New A' of 1889, and the AA of 1900–03, numbering fifteen and twenty engines respectively. The main difference between the two express types was the boiler capacity, being larger in the case of the later engines, which were the last four-coupled type built for the Victorian Railways.

Meanwhile, freight traffic was being catered for by further batches of inside-cylindered 0-6-0s: five 'Belgian R' class by the Societé St. Léonard in 1883, of thoroughly British design but strange continental ancestry; fifteen X class, corresponding to the D class 4-4-0, and of similar building dates; twenty-five 'New R' class of 1889–91; and thirty-one Y class, with larger fireboxes, in 1888–89. None of these included any features of particular interest, but the last two series had side-window cabs, giving them an

The maid-of-all-work in Victoria was the Class D light 4-6-0 in its various guises. Preserved D3 639 is shown here at Sunbury with an excursion in 1971. (*A. E. Durrant*)

appearance rather similar to some of the North Eastern Railway 0-6-0s in England.

For suburban work, a total of seventy-one 2-4-2T with inside cylinders, class E, was turned out by various firms between 1888 and 1894, and these were joined in 1910 by seven rebuilt to the same wheel arrangement from the old Beyer Peacock 2-4-0s. The E class, in turn, became the object of rebuilding, and after five 0-6-2T version of the original E class (EE class as modified) were built in 1893, another twenty-one 2-4-2T were rebuilt to the six-coupled type, giving more adhesion weight.

Around the turn of the century, 2ft 6in gauge branches were built in country where broad gauge lines might prove too costly to construct, and the first two narrow gauge engines were built by Baldwin in 1898. Typical of American narrow gauge power, they had outside bar frames and outside cylinders. Classed NA, the first was a simple engine, but the second featured Vauclain superimposed compound cylinders. A very successful little design, fifteen more were built in the railway's Newport workshops, one of which was also a Vauclain compound. These engines became the standard power on narrow gauge branches, and only one other class, the Garratt, was ever supplied. Today the remaining few operate the 'Puffing Billy' line near Melbourne. With the narrow gauge engines we come to the end oj nineteenth-century Victorian locomotive practice, lagging well behind that of neighbouring New South Wales, particularly in respect of the size and power of locomotives used, notwithstanding Victoria's wider gauge.

However, Victorian practice took a definite turn for the better in the twentieth century, and a really steady programme of development was pursued. Like so many railways, the change in technique coincided almost exactly with the change in century, and in Victoria two new locomotive classes appeared in 1900. One of these, the AA 4-4-0, was the last fling of nineteenth-century practice, as already mentioned. The other, the V class 2-8-0, was the last purely American design to appear in Victoria, but it was also the first eight-coupled type in the state, and as such was a precursor of the freight power which was still some two decades away. The V class as built were Vauclain compounds by Baldwin, and like most of the freight Vauclains with small wheels, clearance considerations dictated that the low-pressure cylinders were positioned above the smaller high-pressure ones. As was usual with Vauclain engines, uneven stresses and wear on the crossheads caused maintenance problems, and they were converted to simple expansion in 1912–13.

The first result of twentieth-century thinking came in the form of the DD class 4-6-0, a solid, simple and dependable engine suitable for both freight and passenger work, of which the first emerged from Newport works in 1902. This was the same year in which Churchward on the Great Western Railway in England introduced his remarkable 4-6-0 with modern front end arrangements, and although the Victorian engine was less technically advanced, it was a substantial improvement over the power previously available for hauling VR trains. Inspiration for the DD must have come from the neighbouring states of NSW and South Australia, both of whom had previously introduced similar types of 4-6-0. The Victorian engines formed a half-way stage in size, and a useful comparison may be made between the three types of broad and standard gauge Australian 4-6-0:

Railway	*South Australian*	*New South Wales*	*Victorian*
Engine class	R	P	DD
Date introduced	1886	1891	1902
Cylinders (in)	18 × 24	20 × 26	18 × 26
Wheels	4ft 6in	5ft 0in	5ft 1in
Grate area (sq ft)	17·6	27·5	22·5
Pressure (lb/sq in)	145	160	175
Tractive effort at 85% (lb)	17,700	23,600	21,600
Locomotive weight (tons)	39·9	56·75	51·5

The DD class proved so useful and popular that 261 were built by 1920, making them by far the most numerous class on the system, amounting to about 40 per cent of the locomotive stock. With so many engines, detail variations occurred; larger boilers, larger cylinders, and superheaters were all applied, not all as a single redesign, so that eventually the class was subdivided; D1 covering the original saturated design, D2 those superheated with original boiler shell, and D3 those rebuilt with new larger boilers based on the K class 2-8-0. Externally, some of the early DD class had a low running plate, gracefully curved over the leading coupled axle to reach a higher elevation over the inclined cylinders, but most were more modern in having a running plate above coupled wheel level, thus eliminating the need for splashers. Light enough to use on many branches, yet powerful for main line work, the various D series 4-6-0 lasted well into the later days of steam; one D3 was still in working order at the time of writing, and used for excursion traffic.

For the heaviest suburban traffic, a 4-6-2T version of the DD class was built from 1908 to 1913, to a total of fifty-eight engines, so that the tank and tender versions of the same basic design amounted to about half the locomotive stock. There can be few railways in the world where at the height of steam operation so large a proportion of the motive power was to one basic design. From 1919 onwards, electrification of the Melbourne suburban services gradually ousted steam power, and two of the tank engines, class DDE, were converted to tender engines, in a similar manner to the NSW conversion of 30 class to 30T after the Sydney electrification. Surprisingly, Victoria did not continue this process, and most of the tanks, later reclassed D4, remained unconverted, relegated to branch and shunting

The first eight-coupled locomotives in Victoria were these Vauclain compound 2-8-0s by Baldwin and Phoenix in 1900–1903. Like most of their breed, they were eventually converted to simples. (*VR*)

Left: Victoria's most modern suburban motive power was the classic but dated D4 4-6-2T, represented here by No 268, the museum example. (*A. E. Durrant*)

Below: An aesthetic paradox, the VR Class A series of 4-6-0 took many forms, but looked best as first built or finally modified. No 964 represents the final form with 'basher' front end, smoke deflectors, and Boxpok driving wheels, seen here as a public park exhibit. (*A. E. Durrant*)

work. Three were superheated, making them tank versions of the D2 class, whereas most were tank equivalents of the D1s.

With the success of the DD 4-6-0s it is not surprising to find that the VR enlarged upon the theme, producing an engine with larger wheels for express work, and larger boiler to permit long periods of heavy steaming. The A2 class as it was designated, first appeared in 1907, at which time it was by far the biggest express engine in Australia, in fact the only six-coupled type with wheels large enough to be classed an express engine. Thow's P class in NSW was neither so large, nor was it an express engine. The original A class design comprised 125 engines, with inside valve gear and outside cylinders, and from 1915 to 1922 sixty more with superheaters were built with outside Walschaerts gear actuating piston valves above the cylinders. Many of the earlier type were later superheated. The saturated engines became class A1, and the superheated examples of both types A2, although it would have seemed worthwhile to distinguish the Walschaerts engines by calling them A3s. For the remainder of the steam era the A2s were the mainstay of VR's main line passenger services, the subsequent S class Pacifics being too few, and the R class Hudsons too late to have much effect on A2 operation. With 185 engines in the class, the A2s were also used on all manner of ordinary passenger and heavy freight duties, and indeed, until the C class 2-8-0 were introduced in 1918, these essentially express engines were also the most powerful that the railway could muster for freight trains. Some comparison may be made between the A2 class and the contemporary NSW N, whilst for those familiar with British practice, the nearest equivalent in size and power is the H15 4-6-0 designed by Urie on the London & South Western, this being particularly comparable with the later Walschaert-fitted A2s.

In later years, many A2s were fitted with improved front ends comprising altered draughting, lowered blastpipes and large-diameter stovepipe chimneys, being known as 'basher' front ends. With the reduced back pressure brought about by these devices, smoke drifting problems became acute, and deflecting plates were fitted beside the smokeboxes. In the very last days of their lives, a few even received Boxpok driving wheels, while at other times have seen feed water

Overleaf: The 'Puffing Billy' line in Victoria is a jolly reminder of the narrow gauge. This scene of a narrow gauge train on a curved wooden trestle, set in the scenic Dandenong hills, is marred only by the dangerous practice of juveniles hanging their legs out of the carriage windows. (*A. E. Durrant*)

Railway	*VR*	*NSW*	*VR*	*LSWR*
Class	A (orig.)	N	A2 (WVG)	H15
Date introduced	1907	1909	1915	1913
Cylinders (in)	21 × 26	$21\frac{1}{2}$ × 26	22 × 26	21 × 28
Wheels	6ft 1in	5ft 9in	6ft 1in	6ft 0in
Grate area (sq ft)	32	26·9	32	30
Boiler pressure (lb/sq in)	200	180	185	180
Tractive effort at 85% (lb)	26,650	26,649	27,480	26,200
Locomotive weight (tons)	68·9	67·4	73·0	79·9

heaters, oil firing, and pulverised coal firing. Altogether, in terms of work done, and long life coupled with low maintenance, the A2s and the D series of 4-6-0 were the best investment in motive power ever made by the Victorian Railways.

In 1918, the VR introduced its first modern freight design, in the C class 2-8-0. These were in some ways a freight version of the later Walschaerts-fitted A2 class, but were enlarged all round so that the similarity was more apparent than real in terms of interchangable components. Whereas at that time the A2 class was the largest and most powerful passenger design in Australia, the C class immediately captured the corresponding title for freight engines. The Cs were big engines with a higher tractive effort than all British 2-8-0s, and an axle load equalled only by the Gresley three-cylinder O2, and exceeded by the Churchward 47XX. Their only weak point seemed to be grate area, on the small side compared to the rest of the engine, and restricted by the confines of a narrow firebox between the frames. They were the last Victorian main line engines to be designed with narrow fireboxes, although production of the lightweight K class 2-8-0 continued for some years afterwards. Only twenty-six C class were built, from 1918 to 1926, and they were converted to oil-burning in 1946–49, had 'basher' front ends fitted, and ended up with the post-war type of German 'blinker' smoke deflectors. Fairly early victims of dieselisation, all disappeared by the early 1960s, with C10 preserved.

Whilst the C class was in production, another 2-8-0 was designed and built for branch line duties, generally to replace the old 0-6-0 engines, and to supplement the D series of 4-6-0. These K class engines had smaller wheels than the 4-6-0s, but more tractive effort and adhesion weight, so that they were more effective in hauling heavy wheat loads over severe gradients, whilst the slightly lower speed potential hardly mattered in that kind of service. Ten only were

At one time the Victoria Class C 2-8-0 was the most powerful and the largest freight locomotive in Australia. C16 shown here is fitted with Fuller–Lehigh pulverised coal firing. (*VR*)

The uninspiring Class K 2-8-0 was one of the last types in regular service in Victoria. K184, returning with a special from Queenscliff to Melbourne, makes a pleasant picture at Geelong, with smoke, clerestory coaches, and a signal gantry. *(A. E. Durrant)*

built in 1922–23, after which it was decided to build all new power suitable for conversion to standard gauge, as there seemed at that time a possibility that the broad gauge systems would be converted to 4ft 8½in, giving a desirable uniformity of gauge throughout the most heavily populated corner of the country. As it is easier to change gauge, particularly from a broader to a narrower, with wide-firebox engines having trailing trucks, than with narrow-firebox ones, the decision was made that all new power should be built with wide fireboxes. However, the replacement design for the K class, the N class 2-8-2, was too long for branches equipped only with 53ft turntables, so during the second world war when these branches were pressed for motive power, a further forty-three of what by then were obsolescent locomotives were placed in service. The final batch of these had Boxpok driving wheels. Devoid of modern cylinder and valve design, as indeed were most Victorian locomotives, we find that the very last locomotives at work in the state, admittedly on very intermittent duties, were a handful of K class, still nominally in service at the end of 1972.

The replacement design for the K class was the N class 2-8-2, incorporating K class running gear with a larger boiler and firebox supported with a trailing two-wheel truck. Thirty were built in Newport works from 1925 to 1931, after which production lagged for some time. In 1927, one engine was fitted with a Franklin booster, and all were later equipped with 'basher' front ends. After the second world war, further light lines engines were required to replace other ageing machines, and more N class were built in 1949–51, fifty by North British and three at Newport. These differed from the original batch in having

Top: The handsome and powerful Class S 4-6-2, Australia's largest Pacific type. Shown here S300 as built. (*VR*)

Centre: Class S 4-6-2 with modified blastpipe, chimney, and with smoke deflectors added. (*VR*)

Bottom: Class S 4-6-2 No 301 enshrouded in an 'upturned bathtub' and with large 12-wheeled tender, for the 'Spirit of Progress' train. (*VR*)

combustion chambers in the fireboxes, which doubtless improved steaming and efficiency, while the Newport engines had Boxpok wheels as well. Full-depth smoke deflectors were applied to the Scottish batch, and German 'blinkers' to the locally-built locomotives. Shortly after delivery of these engines, ten of the Scottish batch were sold to South Australia. The arrival of the later

J class for similar duties meant an early demise for the N class, the last of which was retired in 1966.

Following the N class, the next new type in order of construction was the G class 2-6-0+0-6-2 Garratts built for the 2ft 6in gauge sections. These engines were described in this author's *The Garratt Locomotive*, and there is little to add other than to say that the engine at the 'Puffing Billy' museum at Menzies Creek is not in working order due to the boiler being worn out. Various schemes have been mooted towards the attractive idea of running this engine, but so far none have proved practicable.

1928 saw the first new express passenger class to run in Victoria for thirty years, and these were as great an advance on the A2 class as were these latter over the preceding 4-4-0s. Classified S, the new Pacifics were the largest passenger engines to run in Victoria, and they vied with the South Australian 600 class as to which was the largest express engine in Australia. The S class had rather more tractive effort, and the 600 a slightly bigger boiler, and larger driving wheels, so that the two engines were very closely matched indeed, and had almost exactly the same maximum axle load. The S class had three-cylinder propulsion for the first time in Australia, and they were both in concept and appearance an enlarged version of Gresley's A1 (later A10) Pacific, complete with that designer's conjugated motion for the centre valve. Only four were built, as their heavy axle loading precluded use anywhere but on the main line to Albury (for Sydney), and as the depression was well under way in 1933 when the fourth and last example was built, there was never a good case for multiplying the class. They had cast-steel frames in place of the plate frames previously standard, and No S300, like Gresley's original Pacifics, suffered in performance and efficiency due to the use of short lap and travel piston valves. The other three had 6in travel valves to which No S300 was later altered. Whereas Gresley was able to borrow a Castle from the Great Western Railway to show how to design front ends, the only engines closely available to Victoria were the South Australian 600s, which were not so different. Hence, the S class locomotives soldiered on and were modified twice, first of all with a 'basher' front end and smoke deflectors, in which condition they were probably at their most handsome. Unfortunately, they arrived just before the age of streamlining and shortly afterwards, during the period when everything from toothbrushes onwards had to be 'streamlined', the S class were each entombed in a casing resembling an upturned bathtub, given twelve-wheel high-capacity tenders, and painted blue to match the new 'Spirit of Progress' train operating between Melbourne and Albury. There would have been more progress evident had the gauge been unified between Sydney and Melbourne, but bright and shiny new toys appeal to politicians and public alike, and are much cheaper than doing something really worthwhile. In their streamlined form they were given names of historical personages. They covered the 192 miles from Melbourne to Albury non-stop in 3 hours 40 minutes, at an average speed of 52·3mph, not very exciting for a generally level route, but today, with the 'benefits' of diesel traction, the 198-mile journey (via Sunshine) is only eight minutes faster, in other words it still takes 3 hours 32 minutes. The S class was scrapped in the early 1950s and none was preserved for posterity.

Following close on the heels of the S class was the X class 2-8-2 for heavy freight work. Basically, the X was the development of the C class 2-8-0, but fitted with a shortened S class boiler, with the grate area reduced, seemingly an unnecessary capital economy, for the S and X classes could have been more closely standardised. Nevertheless, just as the C class was larger and more powerful than any British freight locomotive of its day, so the X class was larger again, corresponding fairly closely in size and output with the two Gresley P1 2-8-2s. However much they were ahead of contemporary British freight practice, the X class was third down the list in Australia, for by the time they first appeared in 1929, both the NSW and South Australian systems had far larger and more powerful eight-coupled engines in service. As built, the X class had boilers with round-top fireboxes, as in the S class, but as these needed replacement, Belpaire boilers with combustion chambers were fitted, and with 'basher' front ends and smoke deflectors the locomotives became very impressive-looking indeed. Altogether twenty-nine Xs were built at various dates between 1929 and 1947, and all but two had boosters fitted to increase their starting tractive effort. Like the other big steam power in Victoria, they were early victims of the diesel disease, and the last was withdrawn in 1961.

Victoria's largest steam locomotive was the solitary H class 4-8-4, built at Newport in 1941.

Australia's largest steam locomotive. VR Class H No 220 dominates the other exhibits at North Williamstown museum.

(*A. E. Durrant*)

Designed particularly with heavy passenger work in view, the coupled wheel diameter placed it more in the mixed-traffic category, and in fact it was used mainly for freight haulage. The intended service for H220, known as 'Heavy Harry', was the overnight express between Melbourne and Adelaide, to and from the South Australian border. At the time it was built, engineering works on the line were insufficiently strong to bear the weight of the H class, and due to war conditions, no work was possible to remedy this situation. Accordingly, two similar locomotives were not completed, and 'Heavy Harry' found itself working freight on the main line to Albury, and occasionally putting in a little passenger work when no S class was available. Included in these passenger runs were trips on the 'Spirit of Progress', whose fifty-odd miles an hour must have been easy work for a large-wheeled 4-8-4. The Melbourne–Adelaide overnight trains were a much heavier proposition, due to the load of sleeping cars, and the gradients encountered en route. H220 was designed to equal the two A2 4-6-0s normally used on the train, and in all gross capacities it did in fact achieve this result. The usual sad tale now has to be told. Instead of building more H class after the war, diesels were acquired, and as a result the night express, officially entitled 'The Overland', became better known as the 'overdue' train, so appalling did timekeeping become.

Mechanically, the H was remarkable in a number of ways. Firstly, it was undoubtedly the largest non-articulated steam locomotive to run in Australia, was little inferior to the NSW 60 class Garratts in tractive effort, and well exceeded these in steaming capacity and horsepower. On a worldwide basis the H was one of a very rare breed in being a three-cylinder simple 4-8-4. To this author's knowledge, there were only two other engines of this type in the whole world, these being the pair of German 06 class streamlined express engines. As these disappeared soon after Hitler's downfall, H220 was

The only 'Hudsons' in Australia were the VR Class R, fine locomotives which were premature victims of the diesel. R707, retained for excursion work, speeds through Macedon with a train from Woodend in 1971.

(*A. E. Durrant*)

for most of its life a unique locomotive, a fact which possibly few Australians appreciate. A Belpaire firebox with combustion chamber and thermic syphons ensured high steaming rates, and the front end was characterised by a double blastpipe and chimney, the only such installation in Australia. Machinery could have been improved—the bolted frame structure for a locomotive of this size and power would have been better as a cast steel bed, as on the NSW 57 class, and long lap valves would have improved economy and performance at speed. Rather surprisingly, the big-ends had split brasses, an archaic feature requiring constant attention, which brings to mind the heavy broad gauge engines in Spain. Cylinder design was good, with straight ports, but the conjugated gear for the centre valve, with its long shafts subject to torsional strains, was exactly the same type that Gresley abandoned after his first three-cylinder engine and the Germans (who first applied the device) also abandoned after the numerous $G12^{1}$ and $G8^{3}$ world war one freight engines. In this respect, H220 was also remarkable in being the last engine ever built with the first type of conjugated valve gear for three-cylinder locomotives. Although taken out of service in 1958, the Victorians fortunately were loth to scrap their largest and most powerful steam locomotive, and it now rests as a major exhibit in the museum at North Williamstown.

The war precluded the introduction of any further new classes, but already there had been thoughts of a lightweight Pacific to replace the ageing A2 class. A two-cylinder engine was decided upon, rather than repeat the more complicated three-cylinder arrangement as used on the larger H and S classes, and in many respects the new Pacific could be considered a passenger version of the X class 2-8-2, both the boiler design and axle loading being closely similar. War delayed construction of these engines, and with deteriorating coal quality afterwards it was decided to provide mechanical stokers. These increased weight at the rear end such that a four-wheel trailing truck had to be incorporated into the design, making the engines Australia's first and only 4-6-4s, or Hudsons. With Newport works overloaded in making up wartime arrears of maintenance the final steam classes were imported, and the new Hudsons, designated R class, were all ordered from the North British Locomotive Company, Glasgow, seventy being built in 1951–52. One of the remarkable features of the R class was the incredibly close similarity between it and British Railways Britannia class, also introduced in 1951. The main differences were in the R's mechanical stoker and four-wheeled trailing truck, and that larger cylinders were fitted as boiler pressure was lower than the Britannias, but the net result in the form of rated tractive effort was almost identical. Both designs had Belpaire fireboxes, and both were designed to replace large 4-6-0s. In view of the Britannias' prior appearance, it is tempting to think that the R was Britannia-inspired, but the time interval was too small. Moreover the design was evolved by the VR locomotive drawing office staff, under the CME, A. C. Ahlston, and North British must have been well under way with construction of the R class when Crewe turned out the first Britannia. Nevertheless, a comparison between the two types is most revealing, and the relevant figures are tabulated below. Incidentally, by another strange coincidence, the number series chosen for the two classes of locomotives were very similar; during the Festival of Britain in 1951, R704 was exhibited in Glasgow, and Britannia 70004 shown at London's South Bank exhibition.

Railway	*Victorian*	*British*
Gauge	5ft 3in	4ft $8\frac{1}{2}$in
Class	R	7
Type	4-6-4	4-6-2
Coupled wheels (diameter)	6ft 1in	6ft 2in
Grate area (sq ft)	42	42
Tractive effort at 85% (lb)	32,080	32,150
Two cylinders (in)	$21\frac{1}{2} \times 28$	20×28
Boiler pressure (lb/sq in)	210	250
Locomotive weight (tons)	107·6	94
Adhesion weight (tons)	58·5	60·75
Maximum axle load (tons)	19·5	20·25

Like the Britannias, the R class was destined to have an artificially short life, due to each railway accepting the claims of diesel traction. The R class perhaps fared worse than the Britannias in this respect, putting in far less regular service, and especially being taken off express passenger trains at a very early date. For most of their waning years, the Rs were little more than standby power, some being steamed each year to help move the wheat harvest, heavy slow-speed slogging work for which a Hudson was far from being suited. There never were enough express trains in Victoria to justify seventy Hudsons, and with the wisdom of hindsight it does seem that a 4-8-2 or a 2-8-4 with smaller wheels and more adhesion weight would have been a wiser choice. Fortunately, several R class remain but since 1975 none have been serviceable. No 704 is on show at the North Williamstown museum, bearing a special number plate commemorating its previous exhibition at Glasgow. With their small German 'blinker'-type smoke deflectors painted red, matching the running plate valancing, the Rs were amongst the finest looking engines in Australia.

The last steam class to appear in Victoria was the J class light 2-8-0 for work on the wheatlands branches. These were of the same nominal dimensions as the previous K and N classes of 2-8-0 and 2-8-2 wheel arrangements, and came about due to the N class being too long to fit onto the shorter turntables. As the K class was not convertible to standard gauge due to their narrow fireboxes between the frames, the J class had wide fireboxes placed over the frames and trailing coupled wheels, for the first and only time in Australia. This had long been a common practice on the continent of Europe, but the J class was probably influenced in this respect by the Austerity 2-10-0s built a few years earlier in Britain. Like the R class, the Js had Belpaire fireboxes, but without the combustion chamber found on the bigger engine. Also, where the R class had long-lap, long-travel, piston valves, the Js retained the short-lap valves of their predecessors, although new design cylinders with straight ports were provided. Both the J and R classes used SCOA-P wheel centres, with spokes of U-shaped cross section, lighter in relation to their strength than conventional solid spokes. This was an Australian design, and Victoria the biggest user, Tasmania having a few engines with these wheels, whose only application outside Australia seems to have been the Spanish 4-8-4s and Gold Coast 4-8-2s. In Australia, the spokes were arranged with the twin edges of the U facing outwards, inviting comment, but in Spain they were the opposite way round, looking like thick clumsy conventional spokes. The J class came from Vulcan Foundry in 1953–54, and numbered sixty engines, half of which were oil burners. Like the R class they arrived too late; the diesel was established, and many of the class were scrapped without even having a major overhaul, thus compounding the waste of taxpayers' money—new steam engines were scrapped at an uneconomically early date and replaced by diesels of vastly greater cost, to do the same job.

The last years of steam in Victoria were a sad period in which new and reasonably effective locomotives were neglected and run into the ground, and by 1970 there were no steam engines in regular line service, although a few J and K class survived as works shunters, and even made sporadic sorties on pick-up freights.

VICTORIAN RAILWAYS: PRINCIPAL DIMENSIONS OF SELECTED CLASSES

Class	*Type*	*Cylinders in×in*	*Wheels ft in*	*Heating Surface sq ft*	*Super-heating Surface sq ft*	*Grate Area sq ft*	*Weight tons*	*Pressure lb/sq in*	*Tractive Effort at 85% (lb)*
A	4-4-0	18 ×26	6 0	1221	—	18·3	39·6	140	13,900
B	2-4-0	16 ×24	6 0	1015	—	15·3	—	125	9,080
C	4-4-0T	15 ×22	5 0	761	—	10·5	37·9	130	9,130
D	4-4-0	17 ×26	5 0	1054	—	17·8	39·7	130	13,850
E	2-4-2T	17 ×26	5 0	971	—	17·8	50·3	140	14,850
L	2-4-0ST	16 ×22	5 0	1173	—	14·0	37·4	130	10,400
M	4-4-0T	17 ×20	5 0	1025	—	16·5	42·5	140	11,420
N	2-4-0T	14 ×22	5 0	732	—	12·3	29·6	130	7,950
R	0-6-0	17 ×26	4 6	1054	—	17·8	37·7	140	15,200
S	4-6-0	15 ×24	4 0	—	—	—	—	130	12,400
X	0-6-0	18 ×26	5 0	1361	—	21·7	41·5	140	16,600
Y	0-6-0	18 ×26	4 6	1151	—	21·0	40·5	140	18,550
New V	2-8-0	(2)13 ×24 (2)22 ×24	4 8	—	—	—	—	200	—
AA	4-4-0	19 ×26	6 0	1435	—	21·0	48·8	185	20,750
AA	4-4-0	19 ×26	6 0	1170	216	23·0	51·8	175	19,650
ME	4-4-2T	18 ×26	5 0	1025	—	16·5	54·0	140	16,750
A1	4-6-0	21 ×26	6 1	2210	—	32·0	68·9	200	26,700
A2	4-6-0	21 ×26	6 1	2048	375	32·0	73·0	200	26,700
A2	4-6-0	22 ×26	6 1	1709	331	29·0	72·4	185	27,480
D1	4-6-0	18 ×26	5 1	1381	—	22·5	51·5	175	20,500
D3	4-6-0	19 ×26	5 1	1369	263	25·0	57.3	170	22,300
DDE	4-6-2T	18 ×26	5 1	1408	—	22·5	69·0	185	21,700
K	2-8-0	20 ×26	4 7	1442	238	25·8	62·4	175	28,650
N	2-8-2	20 ×26	4 7	1453	324	31·0	76·0	175	28,650
J	2-8-0	20 ×26	4 7	1444	238	31·0	67·0	175	28,650
C	2-8-0	22 ×28	5 0	2090	338	32·0	81·5	200	38,400
X	2-8-2	22 ×28	5 1	2615	492	42·0	103·0	205	39,360
R	4-6-4	21½× 28	6 1	2243	462	42·0	107·6	210	32,080
S	4-6-2	(3) 20½× 28	6 0	3153	570	50·0	114·5	200	41,670
H	4-8-4	(3)21½× 28	5 7	3980	780	68·0	146·5	220	55,000
*NA	2-6-2T	13 ×18	3 0	524	—	9·0	34·4	180	12,900
* G	2-6-0 +0-6-2	(4)13¼× 18	3 0	1054	180	22·6	70·2	180	27,630

*2ft 6in gauge.

CHAPTER 4

SOUTH AUSTRALIA

South Australia compounded the usual Australian gauge malaise by using two gauges on its state railway system. The main network was on the broad (5ft 3in) gauge. Whereas other states such as Victoria, used minor feeder lines of narrower gauge, the narrow (3ft 6in) gauge system in South Australia amounted to over 40 per cent of the state network, with a similar proportion of its motive power. Not only that, but until commissioner Webb introduced heavy power on the broad gauge in 1926 the narrow gauge engines were comparable in power with their broad gauge sisters. Broad gauge rails dominated Adelaide, the capital, and spread east to join up with the Victorian system, thus making Victoria and South Australia the only two adjoining states with a common gauge. The broad gauge also extended north to Port Pirie, Gladstone and Terowie, but further north than these three points narrow gauge ruled, as it did on the isolated system serving the Eyre peninsula.

Locomotives were classed simply by a letter code, starting at A for the broad gauge engines, and working backwards from Z on the narrow gauge. When Webb arrived from America and proceeded to reorganise the railway and its motive power, new locomotives were numbered in hundreds, and the first engine used as the class designation, eg 500 class. Early locomotives had few features of interest, and from 1856 to 1875 the only engines used were a miscellaneous collection of light, four-coupled machines all of 2-4-0 or 4-4-0 types, with tank and tender varieties of each. Some had inside cylinders, and others outside, and they might be compared in non-standard feebleness with the engines inherited by the Cambrian Railway in Central Wales. The first six-coupled engines arrived in 1874, and were inside cylinder 0-6-0s of typical English design by Beyer Peacock, class J. Apparently such huge power was treated with caution, and the next batch of six-coupled engines

South Australia's Class K 0-6-4T preceded and inspired the better-known Irish version, by the same builders, for the same gauge.

(*J. Buckland*)

The light Class Q 4-4-0 was typical South Australian Railways motive power before the Shea regime; represented here by No 86. (*SAR*)

built, the K class 0-6-4T, were much smaller in conception, being Beyer Peacock tank versions of the old Stephenson long boiler 0-6-0. These strange engines seemed popular, and eighteen of them were built between 1879 and 1885, all by Beyer Peacock. In 1882, the same firm built some almost identical engines with larger wheels for the Sligo, Leitrim & Northern Counties Railway in Ireland, where they again proved popular to the extent of becoming that railway's standard, with later developments on the same basic design being built as late as 1949. On both the Australian and Irish systems, the type attained old age, and some of the South Australian examples lasted until 1956.

The first major development in the direction of adequate main line power came in 1881, when Baldwins built two each of a 4-6-0 passenger design, class N, and a 2-8-0 freight, class O. These were of thoroughly American tradition, differing only in having buffing and drawgear of

An early South Australian 2-4-0T for suburban work. Class P No 122. (*SAR*)

English pattern, and were complete with cow catchers, diamond stacks, and enormous headlamps. In 1904 the two freight engines were withdrawn, but the passenger locomotives were each rebuilt to resemble, externally at least, the later Rx 4-6-0s of local design. Apparently neither of the Baldwin designs proved of much use, and even though they were the largest engines on the system, some of the older and smaller machines were considered more satisfactory.

Larger power was definitely needed for the growing traffic, and in 1886, Dübs of Glasgow delivered the first five of what was to become the backbone of broad gauge motive power in South Australia. These were the R class 4-6-0s, with outside cylinders and inside valve gear, similar to but preceding the NSW P class and the Victorian Ds. They proved immediately successful, and the six Scottish-built engines were followed by two dozen built by the Adelaide firm of James Martin & Co, the first of this batch being also the first locomotive constructed in South Australia. From 1899 they were rebuilt with larger Belpaire boilers carrying a higher pressure, and were classed Rx. Between 1909 and 1915 fifty-four of the Rx class were built new by various local and overseas firms, and with the rebuilt R class they became the most numerous class of locomotive to run in South Australia. Good though the R class may have been in 1886, the Rx were sadly lacking in power for conditions in 1915, when the last were turned out, and the close of the first world war saw the system sadly deficient in motive power.

On the passenger side, even smaller engines prevailed, and the twenty-two Q class 4-4-0 by Dübs and Martin, built in 1885 and 1891–92, with small driving wheels, hardly sufficed for the haulage of heavy or fast trains. The eighteen S class of 1894–1904 were certainly faster machines, and at 6ft 6in had the largest coupled wheels used in Australia, the design again being of the inside-cylinder 4-4-0 type of English (or perhaps, in view of the gauge) Irish standards. Whilst the S class locomotives were fast, they were feeble, and double-heading was rife, particularly on the Melbourne–Adelaide overnight expresses. So far, we have dealt with broad gauge locomotives alone and the narrow gauge will be considered later, but in 1922–23 to meet a motive power shortage, five of the narrow gauge T class 4-8-0 were rebuilt to broad gauge and classed Tx. Whilst on the broad gauge they were superheated, and only regained their original gauge in 1949.

In the suburban area of Adelaide, we first have to chronicle the takeover from the Adelaide–Glenelg Railway of a number of jolly little tank engines. There was already a G class 2-4-0T on the SAR, a sort of Isle of Man engine stretched out to broad gauge, and as the Glenelg line had three the same, they also became class G. The other Glenelg engines comprised two 0-4-4T, and four types of 4-4-0T, five different classes in eight locomotives, and these were tacked onto the G class as classes Ga to Ge. Beyer Peacock built six of their inside-cylinder 2-4-0T design in 1884, and Martins continued with another fourteen in 1893, making twenty of these useful little suburban engines, very similar to the contemporary F class in NSW, the South Australian engines being classed P.

With Adelaide's suburban traffic increasing in weight, a far larger engine was designed and built, for the first time entirely by South Australian builders, and these were the second F class, a 4-6-2T with outside cylinders, generally as a tank version of the R class, but with larger wheels and smaller cylinders, giving less tractive effort. These engines, known as 'Dollies', gave very good service on the Adelaide suburban trains, forty-three being built up to 1922. They handled this traffic until diesel railcars came in the 1960s, and the last was not withdrawn until 1968. More might have been built, but with the electrification of Melbourne's suburban service in neighbouring Victoria some of their E class 2-4-2T were rendered redundant, to be bought by South Australia, presumably at bargain prices. They did not prove popular, and only lasted a few years, being placed in service between 1920 and 1926 and withdrawn from 1923 to 1935.

The Webb era

The introductory story of modern South Australian steam power makes fascinating reading. By the time that the first world war was over, the SAR found itself in the unenviable position of having the least powerful and effective locomotives on any of the standard or broad gauge systems in Australia. The superheated standard goods 2-8-0s in NSW were rated at 33,600lb tractive effort, and the even larger C class in Victoria had 38,400lb, both at 85 per cent boiler pressure. Beside this the 21,420lb of the Rx class was feeble indeed, while the comparison between passenger engines, 27,340lb for Victoria's A2, and 29,000 from the NSW NN class, both 4-6-0s, made South Australia's S class 4-4-0, with its 12,710lb rated tractive effort, almost un-

Light, speedy, but feeble, the SAR Class S 4-4-0 had the largest coupled wheels (6ft 6in) used in Australia. (*J. Buckland*)

believably underpowered. And tractive effort was important, for the climb over the Mount Lofty ranges out from Adelaide involved 1 in 45 gradients uncompensated for ten-chain curves, making an equivalent of 1 in 37, or about the same as the Lickey and Dainton banks in England. In the fascinating *500* book listed in the bibliography, the authors explain how the reigning CME, B. F. Rushton, had struggled from 1913 to 1921 in an attempt to design a locomotive of double the power of the Rx class, only to give up in the end, claiming it was impossible! Exactly what difficulties Rushton encountered or invented are not detailed, but it would seem that either he was not trying very hard, or else his initials were more apt than unfortunate. Curvature, while sharp, was not excessive, and 80lb rail certainly not too light to bear a reasonable-sized engine. In fact, one had only to cross the Indian Ocean to South Africa to see what the other SAR was doing on narrow gauge to dismiss Rushton's efforts as being unbelievably ineffective. Hendrie in South Africa had in 1912 brought out his 12 class with 41,680lb tractive effort (more than any Australian engine at the time), and by 1920 had capped this with the 47,420lb 12A class, both 4-8-2 tender engines. By 1920, Beyer Peacock had built South Africa's pioneer GA class Garratt with 53,700lb tractive effort, and in an article entitled 'The Advantages of Articulated Locomotives' published in *The Railway Gazette* for 28th July 1922, W. Cyril Williams outlined a 4-8-2+2-8-4 Garratt with 81,200lb of tractive effort, and 75sq ft of grate area. This latter was between three and four times the capacity of an Rx, and narrow gauge to boot, yet Rushton was unable to work out a design of half its power!

A merciful veil is drawn over the next moves, but in 1922 we find that the South Australian Railways had a new commissioner, one W. A. Webb, and a new CME in F. J. Shea. It seems that Webb knew what he wanted, and Shea was able and willing to carry out his requirements. Webb was an American, and laid down a series embracing 4-6-2 passenger, 4-8-2 mixed-traffic, and freight designs ranging from 2-8-2 to 2-10-2, the latter with tractive effort specified as 60,000 to 65,000lbs. The requirements were soon watered down somewhat to suit South Australia's track conditions, and quotations called for a 4-6-2 express, 4-8-2 heavy-duty, and 2-8-2 freight types, all to typical contemporary American design with bar frames, big boilers, and two cylinders fed and exhausted through piston valves of ample diameter but inadequate travel and steam lap. When the bids were received, the best offer came from Armstrong Whitworth of Newcastle, with whom the contract was let for thirty engines. Even before they were built, the Webb engines attracted an immense amount of interest, and descriptions, dimensions, and outline diagrams were fed to the main British railway and technical journals, and of course to appropriate Australian periodicals.

In ordering, for 80lb rail, engines with axle

The largest-wheeled Pacifics in Australia were SAR's Class 600, with 6ft 3in driving wheels. One of these massive machines poses in original condition.

(*SAR*)

loads ranging up to 24 tons, Webb was taking rather a chance, but the gamble paid off and rail breakages apparently were limited mainly to occasions when these huge engines were run over 60lb and even 50lb rail, in which cases it was hardly surprising. Ten engines of each class were built—the 500 4-8-2s, 600 Pacifics, and 700 2-8-2s, thus instituting a new numbering and classification system to distinguish the locomotives. All three designs eclipsed anything else in Australia, and also in Britain, for in 1926, when Armstrong Whitworth outshopped them from their Scotswood works, the 600s were the first standard or broad gauge Pacifics in Australia, dwarfing the 4-6-0 types previously in use on the heaviest expresses, and substantially exceeding in size the three British Pacifics of Great Western, Great Northern, and North Eastern Railways. The 700 2-8-2 was larger, if by a smaller margin, than Gresley's P1 of the same wheel arrangement, with a bigger numerical impact, Webb ordering ten Mikados for a four-hundred-engine railway, where Gresley built only two for a six-thousand-locomotive line. The 700s had, similarly, a smaller margin over the Victorian C, previously Australia's biggest freight engine. The engine which commanded the most superlatives was the 500 4-8-2 for heavy passenger and freight service. This massive chunk of steam power had nothing in the same category with which to compare it in either Australia or Britain. In Australia nothing, full stop. In Britain, Gresley's Garratt had more tractive effort, and might have been more effective in the limited context of the Mount Lofty climb, but as a main line engine for continuing all the way to the Victorian border, the larger boiler, wheels, and tender capacity of the 500 completely outclassed Gresley's U1 2395.

Once their inevitable teething troubles were over, the Webb engines set a hard pace for Victoria and NSW to follow, although these two states soon followed on with the competing 57 class 4-8-2 in NSW and S class Pacific in Victoria. Nevertheless, neither of the other two southern states had a full range of heavy power, and it was to be many years before NSW produced a comparable express engine, or Victoria an equivalent heavy mixed-traffic locomotive. As there exists a fully documented account of the Webb engines (see bibliography) there is no point in simply repeating this information here, but subsequent developments will be briefly outlined.

To increase yet further the tractive capacity, 2-8-2 706 had a booster fitted in 1927, and this was assembled into a Delta trailing truck, as with its American contemporaries. In the following year, 506 was also fitted with a booster, but in

The largest power to run in South Australia were the 500B 4-8-4s, rebuilt from the Class 500 4-8-2s. 504, the only survivor, but not in running order, is exhibited at the Mile End museum. (*A. E. Durrant*)

this case the new Delta truck had four wheels, making the locomotive Australia's first 4-8-4. No more booster applications were made to the smaller engines, but all the 4-8-2s were rebuilt as 4-8-4s with boosters, and reclassified 500B. Improved blastpipes and chimneys became a feature of all the large locomotives, due to excessive back pressure with the original designs and the 500B class in final condition were able to haul 548 tons up the 1 in 45 gradients at 15mph, at which speed they were exerting about 90 per cent of their rated tractive effort, booster included, leaving very little margin for contingencies. In pursuance of the quest for second-hand American ideas, a quest which ended in disastrous dieselisation, most of the 500Bs were fitted with deep valancing below the running plates, rather as earlier applied on the Southern Pacific Railroad to their 'Daylight' 4-8-4s. In this condition the big engines looked rather clumsy, for the running plate was too low to give balanced proportions for this styling. However, it was far better than the many contemporary attempts to streamline steam power, and two engines remained without the 'maxi-skirts' right to the end.

Neither Webb's express nor his heavy engines were multiplied, due to the limited use that could be made of such large machines in South Australia, but another ten 2-8-2s were built in the Islington workshops in 1928–29, differing very little from the 700 class, these new engines being classed 710. All were at first fitted with boosters, but as these rendered them too heavy for branches laid with 60lb rail, these were later removed. Nevertheless, they were heavy engines for track of this category, and in most parts of the world the civil engineer would have demanded at least 80lb rail for engines with an axle load exceeding 18 tons. Hence it was decided to design a new type of locomotive for use on light lines, and as the booster-fitted Mikados produced the required power output, the obvious step of spreading the weight using a four-wheeled Delta truck became the basis of the new engines. Unfortunately, everybody got so carried away with the new design that it emerged with a much larger boiler than before, making a very impressive locomotive indeed, but with a heavier axle load than ever (nearly 20 tons), such that the 720 class 2-8-4 was soon confined to the same main lines as the 500 and 600 classes. Following the first five, built 1930–31, twelve more were turned out from Islington in 1938–43, and these for some extraordinary reason had the same deep valancing below the running plate as the modified 500B engines. The last of the true Webb

Above: The only 'Berkshires' or 2-8-4s in Australia were the SAR Class 720. Over-designed as branch line power, No 734 represents this essentially main line freight hauler class.

(*SAR*)

Below: South Australia's Class 620 light 4-6-2s appeared powerful by virtue of outsize boiler cladding and *Cock O' The North* front end. 624 is in Mile End museum, but 621 is available for excursion work.

(*A. E. Durrant*)

engines were the ten of class 740, built after the second world war as part of an order destined for China, to be supplied as aid under UNRRA auspices. By the time Clyde Engineering works had completed the engines, China had gone through its communist revolution and the engines were not needed. Accordingly, as it was a South Australian design, albeit adapted to standard gauge, the SAR bought ten of them, which became the 740 class, while the others went to

the Commonwealth Railways (see chapter 5).

Meanwhile, well before the second world war, a severe need was felt for engines which really were light enough to perform on 60lb rail, at reasonable speeds with passenger trains, and not be handicapped by the restrictions imposed upon the 700 and 710 classes of 2-8-2. Shea was still the chief mechanical engineer and after he had been dissuaded from building another 2-8-4 class, he supervised the design of a lightweight Pacific, smaller all round than the hefty 600 class. The new engines, designated 620 class, were much more modern in their valve design, and included long-lap, long-travel piston valves, actuated by Baker valve gear. In this respect, British practice had been followed, at least as far as the valve events were concerned, although in Britain Walschaerts or Stephenson motion was always used. As a result, the 620s were snappier and more efficient than the 600s, and the 500s, but these earlier and larger engines were never altered to conform, even though the changes needed to update them would have been simple and inexpensive. The 620s had multiple-jet blastpipes as built, and with a maximum axle load under 16 tons were particularly useful locomotives, being light enough for branch work, yet sufficiently fast and powerful for most of the main line passenger trains. In many respects they were the equivalent of such 'maids of all work' as the Great Western Hall or LMS Class 5, and their only weakness was in maximum capacity. The most unfortunate of the 620 class was the initial engine which, appearing in 1936, South Australia's state centenary, was made a major exhibit at the Wayville (Adelaide) show. Now this would normally have been a good thing, but in those pre-war years when everything had to be 'streamlined', 620 was shrouded in one of the most ghastly casings that ever insulted the proportions of a well-bred steam locomotive. Wheels and motion were left mercifully alone but the boiler front received a strange treatment with chrome grille, making the unfortunate locomotive look like a cross between a jukebox and an early French motor car. Surmounting this monstrous abomination was a chimney-cover-cum-headlight-casing excresence which anticipated the worst of today's plastics age. Fortunately, the remaining nine engines, built 1936–38, were spared the same treatment, and were

The most modern and versatile passenger locomotives in South Australia were the Class 520 4-8-4s, with shark-nose streamlining based on the Pennsylvania Railroad T1. Not nearly so big as they looked, the 520s were light enough for branch line service. (*A. E. Durrant*)

outshopped with frontal appearance resembling Gresley's *Cock o' the North* prototype. 620 herself was later given plastic surgery to remove the unwanted grillwork. Later again, as the *Cock o' the North* front end failed to deflect smoke, plates were fitted to the smokebox sides of all the class. After diesels had replaced the bigger power, the 620s could still be found doing good work on the light branches for which they were designed, and the last were not taken out of service until 1969.

Following the 620 class came another modern lightweight design, the 520 class 4-8-4s. With the same size driving wheels, and the same modern front end as the 520s, these were undoubtedly the most useful and versatile locomotives to run in South Australia. With an axle load of the same order as the 620s, the new engines were a 'do anything, go anywhere' locomotive, and in passenger service could equal the performance of the 600 class heavy pacifics. Mechanically, they had Walschaerts instead of Baker valve gear, and were given a streamline casing based on that of the Pennsylvania T1 4-4-4-4 engines. Two versions of this shark nose were tried, the original type with blunt and vertical lines being rather clumsy, but the later version with sharper, raked-back nose was most attractive, being one of the very few attempts at steam locomotive streamlining which did not detract from the overall appearance. Altogether, only twelve 520s were built, all at Islington, between 1943 and 1947. More were proposed, but the diesel tide overtook them, as it did a proposal for more modern, heavy 4-8-4s to be known as the 800 class.

Although the 520 class was the final new design built for South Australia, the last to appear on the roster, if we count the 740 class as a slight variation on the original 700s, were ten light 2-8-2s classed 750. These were acquired second-hand from Victoria where, although brand new they were surplus to requirements at the same time as South Australia was short of

South Australia Class 750 2-8-2s, second-hand from Victoria, were the last and lightest Mikado type to be introduced on that system.

(*A. E. Durrant*)

power pending the arrival of the 740s. The 750 class were of the Victorian N class, of North British build, and with their light axle load were also useful on light tracks, but being less versatile than the 520s, they were scrapped somewhat earlier.

Narrow gauge locomotives

As previously mentioned, narrow gauge (3ft 6in) played an important part in South Australian operations, the first sections being from Port Pirie to Gladstone, and further south from Port Wakefield to Blyth. The engines imported to work these lines were Beyer Peacock 2-6-0s of a classic design exported by that firm to various parts of the world, especially to Australia. Actually, there were three main versions of these engines, the earlier with horizontal cylinders and overhead slide valves, based on an earlier Neilson design for Queensland, and the later true Beyer Peacock engines with inclined cylinders, these latter being built in two sizes. The first of the Beyer Peacock horizontal-cylinder engines seem to have been outshopped from Gorton in 1875, for the Cape Government Railways in South Africa, and these were followed by the South Australian U class in 1876. The Western Australian M class of 1875 was also of this type, but of Kitson build. The Beyer Peacock design with inclined cylinders seems to have originated on the South Australian system, whose W class came out in 1878, followed quickly by the Cape engines of 1879 and the Western Australian A class of 1880. Both these earlier types, of different design and appearance, had the same sized wheels and cylinders, and the enlarged version of the second series were again introduced in South Australia, as their Y class, in 1885, as C class in Tasmania the same year, and as G class in Western Australia in 1889. These little engines, neat, simple and trouble-free, suited Australian conditions very well, and were still being turned out at the end of the century. Being a sort of Australian standard, they were exchanged between systems, particularly in wartime, and became popular second-hand power on forestry lines.

Returning to the specifically South Australian side of the story, there were eight of the original U class, all of 1876 vintage, while for shunting and short distances there were eight 0-4-4T class V, the first of which came out from Beyer Peacock in 1877 for the Kingston branch, followed by four more locally-built in 1893. They were not considered successful in line service, which was hardly surprising for an 0-4-4T with outside cylinders. The W class 2-6-0 comprised thirty-six locomotives built from 1878 to 1883 by Beyer Peacock, of which sixteen were rebuilt with higher-pitched boilers carrying more pressure, and reclassified Wx. Larger 2-6-0s, class X, were supplied by Baldwin in 1881–82, and in 1885 the Y class came out from Beyer Peacock. These were to the narrow gauge what the R class was on the broad, an all-purpose engine numerous enough to handle most of the traffic offered. Some 123 were built, the earlier ones from Gorton but most locally-produced by Martins. Fifty-eight were rebuilt on the same lines as the earlier engines, ie with higher-pitched boilers and increased pressure, this time also with Belpaire fireboxes. Strangely, not one of the Beyer Peacock engines was rebuilt to class Yx, although there seems no logical reason for this. Naturally, on these narrow gauge outback lines, most passengers were conveyed by mixed trains, but for the handful of faster passenger trains, particularly the Broken Hill expresses, ten 4-4-0s were built, eight by Martin in 1895 and two at Islington as late as 1911. Despite their local build, these delightful little engines were as Beyer Peacock as Gorton could produce, and were evidently based upon the similar locomotives supplied from Manchester to Japanese railways a little earlier. Possibly Martins purchased the design from Beyer Peacock, but whatever their parentage all were highly successful, and remained in service until 1956.

Before continuing with the two later and most important SAR narrow gauge classes, two species of tank engine claim our attention. First was class K, whether or not by coincidence an inside-cylinder 0-6-4T very similar if slightly smaller than the broad gauge K class. The narrow K was Dübs built in 1884, and although not repeated, managed to last until 1938, being one of the very rare inside-cylinder narrow gauge types, seen hardly anywhere else throughout the world.

The first big power on the narrow gauge lines appeared in 1903, when Islington shops turned out the pioneer T class 4-8-0. With inclined outside cylinders, inside valves and gear, these clearly derived much from the Cape Government Seventh Design in South Africa, but were sufficiently different not to be called a straight copy of the earlier African engines. With their inclined running plates over cylinders and slide bars, there was a bit of Beyer Peacock about the T class, although all were Australian-built. Outside

South Australia Class T 4-8-0s, despite the Beyer Peacock and South African influences in their appearance, were locally designed and built. No T243 is shown on the Wilmington branch in 1969.
(*J. A. Joyce*)

frames on the bogies gave them a rakish air, as did the Belpaire fireboxes, and extended smokeboxes topped originally by tapered capped chimneys but replaced in 1940s/50s with plain stovepipe chimneys which added a pugnatious look quite in keeping with their performance. With the introduction of these 4-8-0s, the narrow gauge possessed engines with as much tractive effort as the rival broad gauge biggest machines, the Rx class, a position which remained until Webb introduced his heavy broad gauge power, over twenty years after the first T class. By the end of 1917, all seventy-eight of the T class had been built, and they henceforth became the standard motive power on the narrow gauge, almost equalling in numbers the broad gauge Rx class. Despite their undoubted worth, it seems surprising that Webb did not see fit to introduce big power for the 3ft 6in sections of the system at the same time as his big broad gauge engines came out. In fact by 1926, the year that the 500, 600 and 700 classes of broad gauge power were built, South Africa had the remarkably similar 15CA and 15CB heavy 4-8-2s in service, which would have been most fitting narrow gauge companions for Webb's 500 class. The above remarks are best illustrated by a table showing principal dimensions of the Cape Seventh and South Australian T class, as a direct comparison, the South African 15CB as what might have been, and the South Australian 500 as a control comparison. The SAR 15CB figures are as built, and not as now running with larger wheels and cylinders.

However, as things were, the T class was to rule the South Australian narrow gauge for fifty years before more powerful locomotives came along, and some remained at work until January 1970, during which month South Australia's last regular steam workings were run, divided be-

Railway	*South African*	*South Australian*	*South African*	*South Australian*
Gauge	3ft 6in	3ft 6in	3ft 6in	5ft 3in
Class	7	T	15CB	500
Introduced	1892	1903	1925	1926
Type	4-8-0	4-8-0	4-8-2	4-8-2
Cylinders (in)	17×23	$16\frac{1}{2} \times 22$	23×28	26×28
Driving wheels	3ft $6\frac{3}{4}$in	3ft 7in	4ft 9in	5ft 3in
Grate area (sq ft)	17·5	17·3	45	66·6
Tractive effort at 85% (lb)	21,900	21,150	44,180	51,000

tween the T class and the 400 Garratts. More will be said about these final workings shortly.

As the T class locomotives approached their half century of service, greater power became necessary, and the South Australian Railways ordered from Beyer Peacock ten 4-8-2+2-8-4 Garratts of the same design as the East Africa 56 class, built in 1949. Gorton was snowed under with work and sub-let the contract to Franco-Belge, but to tide over the period, six of the ASG Australian Standard Garratts were purchased from Western Australia. More will be said about these machines in a subsequent chapter; sufficient to record here that the WAGR was pleased to get rid of engines only ten years old and designed by their own CME, while in South Australia their life span extended to but four years. The six 300 class ASGs hauled iron ore from Cockburn to Port Pirie until the arrival of the 400 class Garratts in 1953, whereupon they were phased out.

The 400 class Garratts were certainly a great

400

The very last regular steam operation in South Australia was by Garratts on the narrow gauge. Shown here in the last months, No 400 leaving Terowie for Peterborough.

(J. A. Joyce)

success, and revolutionized ore workings on the line, being equal to two T class. As finally delivered they were almost identical to the East African 60 class, developed from the 56 class, but actually built after the Australian batch. Even these were not the first built to this design, this distinction belonging to a 1952 Henschel batch for Brazil; further information on this interestingly international Garratt will be found in *The Garratt Locomotive*. As the final steam class ordered and built for South Australia, the 400 class fittingly hauled the very last steam train to run in South Australia, other than enthusiasts' specials. In this author's *The Garratt Locomotive* published in 1969, it was stated that the last news of the 400s gave them as being in stock but out of use due to dieselisation. However, towards the end of 1969, with the narrow gauge scheduled for conversion to standard, the diesels were also withdrawn from service for gauge conversion, and for a brief period of final glory, T class 4-8-0s and 400 Garratts, often double-headed in various combinations, handled the Broken Hill ore traffic between Cockburn and Port Pirie. For those interested in further details of these final workings, *Proceed to Peterborough* is thoroughly recommended reading matter, illustrated with excellent photographs.

SOUTH AUSTRALIA: PRINCIPAL DIMENSIONS OF SELECTED LOCOMOTIVES

Class	*Type*	*Cylinders in × in*	*Wheels ft in*	*Heating Surface sq ft*	*Superheating Surface sq ft*	*Grate Area sq ft*	*Weight tons*	*Pressure lb/sq in*	*Tractive Effort at 85% (lb)*
A	2-4-0T	14 × 22	5 0	817	—	11·2	29·7	130	7,940
B	2-4-0T	14 × 20	5 6	794	—	11·6	31·9	130	6,560
C	2-4-0	14 × 20	4 6	854	—	12·7	25·7	130	8,020
D	4-4-0	15½ × 22	5 4	951	—	14·7	28·3	130	9,130
E	2-4-0T	15 × 21	5 6	895	—	13·0	31·8	130	7,910
E	2-4-0	15 × 21	5 6	895	—	13·0	27·3	130	7,910
F	4-4-0T	14 × 22	4 11½	—	—	—	30·4	—	—
F	4-6-2T	17½ × 24	5 3	1335	—	18·0	59·0	185	18,335
G	2-4-0T	11 × 18	4 0	—	—	—	21·0	130	5,010
Gd	0-4-4T	14 × 20	4 9	603	—	10·5	23·2	145	7,630
Ge	4-4-0T	13 × 18	4 0	636	—	11·4	28·9	145	7,810
H	4-4-0	14 × 20	4 6	854	—	12·7	26·9	130	8,020
J	0-6-0	17 × 24	5 0	1013	—	15·0	30·2	130	12,770
K	0-6-4T	16½ × 20	4 0	920	—	15·0	40·8	130	12,540
L	4-4-0	16 × 22	5 0	972	—	15·6	31·9	130	10,370
M	0-4-2T	12½ × 16	4 7	508	—	8·4	22·5	130	5,020
M	2-4-2T	17 × 26	5 0	1072	—	17·8	53·4	140	14,900
N	4-6-0	19 × 24	5 0	1303	—	19·0	42·8	130	15,900
O	2-8-0	20 × 24	4 0	1336	—	32·7	47·2	130	22,100
P	2-4-0T	16 × 20	5 0	934	—	14·6	32·6	145	10,520
Q	4-4-0	16½ × 24	5 1	1122	—	16·1	32·5	130	11,840
R	4-6-0	18 × 24	4 6	1294	—	17·6	39·9	145	16,800
RX	4-6-0	18 × 24	4 6	1437	—	20·2	45·0	175	21,420
S	4-4-0	18 × 24	6 6	1231	—	17·4	38·4	150	12,710
500	4-8-2	26 × 28	5 3	3648	835	66·6	134·9	200	51,000
500B	4-8-4	26 × 28	5 3	3648	835	66·6	143·8	200	51,000
600	4-6-2	24 × 28	6 3	3283	743	55·0	117·0	200	36,600
620	4-6-2	18½ × 28	5 6	1735	421	33·4	81·5	200	25,000
520	4-8-4	20½ × 28	5 6	2454	651	45·0	111·3	215	32,600
700	2-8-2	22 × 28	4 9	2595	619	47·0	96·2	200	40,400
720	2-8-4	22 × 28	4 9	2975	751	59·5	123·4	215	43,400
* K	0-6-4T	14½ × 20	3 6	777	—	13·1	32·6	130	11,100
* U	2-6-0	12 × 20	3 3	538	—	9·8	18·9	130	8,160
* V	0-4-4T	9½ × 15	3 0	257	—	4·7	15·7	130	4,150
* W	2-6-0	12 × 20	3 3	544	—	9·8	18·5	130	8,160
*WX	2-6-0	12 × 20	3 3	548	—	9·8	20·5	145	9,101
* X	2-6-0	14½ × 18	3 2	563	—	14·6	22·1	130	11 000
* Y	2-6-0	14½ × 20	3 3	778	—	13·7	24·3	130	13,290
*YX	2-6-0	14½ × 20	3 3	806	—	13·7	28·2	185	16,955
* Z	4-4-0	15 × 20	4 6	869	—	13·9	28·3	145	10,270
* T	4-8-0	16½ × 22	3 7	764	136	17·3	41·3	185	21,904
*400	4-8-2 2-8-4	(4) 16 × 24	4 0	1972	370	48·8	149·0	200	43,520

*3ft 6in gauge.

Chapter 5

COMMONWEALTH RAILWAYS

In Australia there were three layers of railway ownership. Firstly there was the handful of privately-owned concerns for hauling bulk traffic, such as coal, ore, and sugar cane. These were usually short on mileage and high in traffic density and either profit-making in their own rights, or indispensible adjuncts to the owning industries. The second and major group comprised those of the individual states, essentially government-owned due to the numerous pioneer lines built to open up land for settlement, lines whose profitability could only be reckoned in state prosperity, and not within the narrow confines of conveying freight at so much per ton. The tertiary layer, unique to Australia, was the Commonwealth Railways organisation, created to run such railways that even the individual states found beyond their means. These served those districts even less populated and more remote than the outback lines of individual state systems, and tended to run from nowhere to nowhere! Even the federally-operated Commonwealth Railways was not free from the country's gauge troubles, and of the four sections of railway owned, the only adjacent two were of different gauges!

The principal CR line was the Trans Australian, a standard gauge connection from Port Augusta, South Australia, to Kalgoorlie, Western Australia, traversing the Nullarbor (no trees) plain, a 1000-mile stretch of virtual desert. It seemed somewhat muddled thinking that built to 4ft 8½in gauge a long and costly railway to connect two 3ft 6in gauge state systems! Had the obvious step been taken, of connecting the two 'Cape gauge' systems with a link of uniform gauge, then the ultimate standardisation of this gauge could, and probably would, have been effected at a fraction of the cost of trying to standardise the broader gauge. It must be realised that 4ft 8½in gauge was chosen with the full concurrence of CR engineers in 1903 and at that time there were no high capacity 3ft 6in gauge railways anywhere in the world (South Africa included). Moreover gauge standardisation in Australia seemed to be a probability, and there were firm plans to carry the standard gauge right through to Perth which was eventually achieved in 1970.

In the matter of motive power the TAR showed amazing commonsense, and instead of ordering complex and untried locomotives straight off the drawing board, existing and well proven designs were chosen. As the only other 4ft 8½in gauge line was that of NSW, then NSW types were chosen, but at the time they were probably the best in Australia, with the possible exception of Victoria, whose broad gauge engines would have needed extensive redesign. Work on the line was commenced from each end in 1912, and late in 1917 the two sections were linked near Ooldea, Australia's equivalent of Promontory.

There is little to say about the locomotives used on the TAR, for on the one hand, none were designed specifically for the railway, and on the other a book, *Locomotives of the Commonwealth Railways*, has just been published giving full details. Hence, all that seems necessary is a brief catalogue listing locomotives used.

The first engines used on the TAR were six 4-4-0, second-hand ex-NSW Q class, rebuilt from Beyer Peacock tank engines (see p. 16). Occupied firstly on construction work, they were later used for shunting, some surviving the 1939–45 war. For main-line work, Thow's well-tried designs from NSW were chosen, and twenty-six passenger 4-6-0s of the NSW P (later 32) class were built in 1914–17, and class G. These had few alterations other than the fitting of headlights, cowcatchers (or, perhaps,' Roo catchers), and buckeye centre couplings. Drumhead extended smokeboxes were fitted, neater than the rather clumsy NSW extensions. All were built saturated, but seven were later superheated and classed GA. An odd Baldwin 2-6-0 ex-NSW became class F, but the main freight types were eight 2-8-0 of the NSW T (50) class, and twenty-six of the TF developments. These became classes K and KA on the CR, and differed from their eastern equivalents in exactly the same details as the passenger engines.

Most imposing of the CR locomotives were

Original motive power on Comrails was their Class G, built to NSWGR Class P drawings. G1 is preserved at Mile End, South Australia. (*A. E. Durrant*)

the eight C class 4-6-0s, equivalent to the NSW 36 class, but with huge twelve-wheel tenders, longer and heavier than the locomotives themselves. During the second world war, with motive power at a premium, the CR was forced to obtain second-hand power from the only source available, that is North America. Ten rather elderly 4-6-0s, approximating in capacity to the C class, were obtained, eight from the Canadian National, and two from the New York, New Haven, & Hartford. These were known in Australia as CN and CA respectively, and were one of the very few instances of North American locomotives sold for use outside their native subcontinent. The final TAR steam class comprised ten 2-8-2s to the South Australian 700 class, built for China but rendered surplus for political reasons (see p. 62). As CR class L, they arrived at the same time as diesels, and some never did a day's work, being cut-up as rusted hulks of brand-new engines.

Serious dieselisation of the CR started in 1951, and was virtually complete two years later. Even this author, steam diehard that he is, can see some logic in dieselising the TAR, running as it does through over a thousand miles of inhospitable terrain, with infrequent supplies of very bad water, and with coal supplies up to two thousand miles from coaling stages. Even so, it is nice to think of what could have been done with steam—had the line been built to the obvious 3ft 6in gauge, something like the South African 25 class condensing 4-8-4s could have been employed, which, at 4,000hp well exceed the 3,300hp output of the latest standard gauge diesels used across the Nullarbor plain.

Narrow gauge

As on the standard gauge, all the 3ft 6in gauge engines of the Commonwealth Railways were either second-hand, or built new to an existing state design. These narrow gauge lines were originally part of the South Australian narrow system, and indeed at that time the Northern Territories were administered by the South Australian Government. The original intention was to extend narrow gauge northward to Darwin, and had the Northern Territory continued to be ruled from Adelaide, this would probably have been effected. However, the Northern Territory passed to the Common-

Commonwealth Railways best steam power was the Class C 4-6-0, to the NSWGR Class 36 design, but with twelve-wheel tenders longer and heavier than the locomotives, for running across the Nullarbor Plain. (*Comrails*)

wealth in 1911, leaving the Palmerston & Pine Creek Railway, later known as the North Australia Railway, stranded in Arnhem Land, running south from Darwin to nowhere. The line from Port Augusta extended its tentacle northwards in a vain attempt to make contact, but eventually reached only as far as Alice Springs. Both these lines passed to the Commonwealth Railways in 1926, who completed the last sections south to Birdum and north to Alice. But Darwin is now never likely to be connected to the rest of Australia's rails' for the North Australia Railway closed on 30 June 1976.

So far as locomotives go, we find that the first engines on the Palmerston & Pine Creek comprised a minute Baldwin 0-4-0ST, and six ex-SAR 2-6-0s of their class W, these becoming Commonwealth classes NA and NF respectively. Class NFA was a Dübs 2-6-0, a solitary engine probably built for main line work somewhere else in the world. Other second-hand power scraped up from assorted sources across Australia were the NG (six 4-6-0, ex-QGR B13), the NFC (thirteen 2-6-0 ex-WAGR G class, some originally from the SAR) and a like number of the 4-6-0 version which were classed NGA. From South Australia came eighteen Yx 2-6-0, as Commonwealth class NFB, and there was an odd 0-6-0ST, and two varieties of steam railcar.

The only narrow gauge power built new for the CR comprised twenty-two 4-8-0 class NM, to the Queensland C17 design. Despite their diminutive proportions, these were too heavy for the Darwin line, and used exclusively on the 'main line' to Alice Springs. Eleven of Queensland's smaller C16 class also worked for a while as class NMB. Commonwealth locomotives are fully documented in a recent book (see bibliography), and with few features worthy of special comment, have been dealt with very briefly.

Commonwealth Land Transport Board—the Australian Standard Garratts

Although nothing to do with the Commonwealth Railways, who wisely declined to employ any, these controversial locomotives are most conveniently dealt with in the Commonwealth chapter. In this author's *The Garratt Locomotive*, these locomotives and their troubles are briefly detailed, but during his Australian sojourn

Principal heavy locomotives of Comrails narrow gauge were the Class NM 4-8-0s, to Queensland's C17 design. NM34 is preserved at Mile End, South Australia. (*A. E. Durrant*)

interesting new facts were accumulated, including a copy of the Royal Commission's report on the engines. This makes fascinating reading, and investigating further, it seems that there are two slightly conflicting versions of which engines were employed where, and whereas all enginemen with first-hand knowledge of the engines as built invariably condemned them, a few *were* made to work in Tasmania and Victoria. There was also a Western Australian opinion to the effect that they must have been good, as they were designed by the WAGR CME, but this may be discounted as misplaced patriotism!

As basic details of the reasons behind the ASG locomotives, and of the numbers built and not built already appear in *The Garratt Locomotive*, it is proposed to comment here mainly on the Royal Commission report. That a Royal Commission was needed to report on something so well established as the steam locomotive invites comment in itself; it would have been surprising in 1846, let alone 1946, especially with plenty of other post-war problems prevailing. It was headed by Mr. Justice Wolff, of the Supreme Court of Western Australia.

The report opens with a preamble, explaining what a Garratt is, expanding to greater detail of other Australian Garratts, and to the ASG itself. Under Part II, *Conduct of Inquiry*, the crux of the problem soon comes to light as a battle between the enginemen's trade union and the 'Department', this latter term boiling down to the WAGR CME, Mills, who designed the engine. Obstinacy was the order of the day, the unions overstated their case, and Mills, like a nineteenth-century British locomotive martinet, refused to concede that anything he or his design staff produced could possibly be tainted with the slightest defect. Had the subject been less specialised and technical, the proceedings might have provided an excellent background upon which to base a comic opera, and the numbered points of contention concerning safety and economy form a useful basis for further comment:

1. *The union complains about the steam brake's unreliability.* This is due to condensation in pipes and cylinders, a very valid point inherent in all steam brakes, and accentuated in a Garratt due to its lengthy piping. The Department asserts that the brake is efficient and reliable, but adds that 'steps are being taken to minimise condensation'. Round one

to the union, the Department evidently taking steps to remedy a fault which they deny exists.

2. *Steam brake power decreases with boiler pressure.* The Department's answer admits this, but adds that 'safety is not impaired'. Round two also to the union, for clearly there comes a pressure below which the steam brake will prove useless. The steam brake does not 'fail safe' and an engine with steam brake and no pressure will be unstoppable. With automatic vacuum brakes, a steam pressure failure will cause the ejector to cease functioning, eventually allowing leaks in the train line to stop the train.
3. *Engines liable to derailment, as leading coupled wheels are flangeless.* The Department's answer was a simple denial, but here again the union was right. Engines with flangeless leading coupled wheels *have* been used successfully elsewhere, but the flangeless tyres have been extra wide, like steamroller wheels, so that they remain on the rails despite throwover on curves. ASGs had normal-width tyres, making them a proven hazard, accentuated by weak sideplay control springs on the bogies. Round three to the union.
4. *Excessive wheelbase and weight of locomotives cause undue stress on trackwork.* This is denied by the Department, and this time they were right.
5. *Excessive width of engine a danger when passing other trains etc.* The Department rightly point out that the engines are within the loading gauge.
6. *The reversing screw needed too many turns from forward to reverse gear.* Here the union grossly overstated its case by claiming that the engine 'often' had to be reversed for emergency stops.
7. *The regulator was insensitive.* This is denied by the Department, which was probably right, as this item disappears from the proceedings.
8. *Steam blows, 'inherent in Garratt type locomotives', interfere seriously with crew's visibility.* As the WAGR had been running Garratts for over twenty years the defect was hardly inherent to Garratts, the union again overstating its case. The Department on the other hand completely denies the allegation, so far as Garratts in general were concerned, but admits the trouble occurred on ASGs.
9. *Visibility fore and aft were restricted.* The Department flatly denies this, but the high pitched front tank, devoid of inboard chamfering, would seem to affect visibility.
10. *Again deals with regulator insensitivity,* and unnecessarily repeats item 7.
11. *Continues with brake troubles,* largely repeating paragraphs 1 and 2, to which the Department answers that brake operation is 'satisfactory'.
12. *When the train is coasting, the weight of the engine causes 'jerks and bumps' on the drawgear.* The Department, quite rightly, deny this. The union then continues under a heading *The Engine is Unweildy and Irksome to Operate.*

Here they claim that operation of levers 'operated perhaps 100 times a day' . . . 'almost all require exertion of strength, and the use of two hands and sometimes a foot'. The old bogey of having 'two engines' and double the number of oiling points also comes up, despite the fact that extra preparation time was allowed for Garratts. There is also the emotional idea that the fireman had to fire 'for two engines' with resultant excessive fatigue. As the ASG was a very modestly-sized Garratt, with only 35 square feet of grate area, there is no justification for the complaint. In fact, the same railway's S class 4-8-2, introduced the same time, had bigger hand-fired boilers than the ASGs! Again, a fireman's complaint referred to having to shovel coal forward at the ends of runs, but the ASG' bunker was smaller in capacity and better designed from a self-trimming aspect than that of the S class, so the complaint was more in the mind than the reality. Although it never says so, the union is clearly worried about redundancies due to 'one man having to do the work of two', although they never complained of larger conventional locomotives being introduced. The Department simply disagrees with this general moan.

Under the third heading, *The Engine is Uneconomical,* the union is on weaker ground. Its points are:

1. That the 119-ton weight of the ASG represents a high deadweight to be deducted from the useful work done by the engine. The Department points out that weight is necessarily proportional to tractive effort, and could have gone further to note that the ASG weighed exactly the same as an S class with tender, which latter engine had less tractive effort.

2. A rather garbled complaint, amongst which is the statement that the ASG does not use full power when hauling empty wagons which, as the Department points out, applies equally to other locomotives.
3. The union complains that a Garratt is uneconomical because it has four cylinders! Chapelon or Churchward would have disagreed with this, and the Department simply counter-claims that one boiler is more economical than two, doing the same work.
4. The union complains that the long steampipes result in loss of superheat, and thereby loss in economy. The Department replies that this is a 'statement of opinion' and disagrees. However, such a phenomenon undoubtedly existed with Garratts, although in a warm climate like Western Australia the effect was minimised. Garratts have been run successfully in climates including snow and ice with no apparent ill effect, and compensation may always be made by increasing the initial superheat. This was not done on the ASG, which had a smaller proportion of superheat to evaporating surface than the S class. Here the union was right, but exaggerated the matter while the Department, which should have had facilities to confirm or refute the claim, simply disagreed.
5. Components frequently break or become defective. The Department disagrees again, but as we shall later see there was a great deal of substance to this complaint.
6. Poor steaming. The Department denied this, of course, but the union was right, as subsequent tests showed.
7. Largely a repetition of 5.
8. The ASG uses twice as much oil as any other class. Not surprising, as it has twice the number of working parts, but the Department merely counters that oil consumption is 'not excessive'.
9. Driver and firemen are allowed 30 minutes extra to prepare, and 15 minutes extra to stable the engines, compared with a conventional engine. These figures were evidently agreed by the Department under union pressure, for the reply agrees that these extra times are allowed but does not concede that they are 'reasonable'. The union's claim that the extra time which they have demanded makes the Garratt uneconomical does not hold water—clearly a little extra time spent by one crew costs less than employing two crews to run a double-header, each with a full allowance of preparation time. Equally clearly, despite the Department's allegation, a Garratt with its extra running gear does need some additional preparation time.
10. The 'unweildiness' of the Garratt causes bumps and damage to other rolling stock. The Department's answer that this was due to mishandling probably approached the truth more closely—much larger engines than the ASG have been successfully driven elsewhere in Australia and the rest of the world.

Part III of the report deals with safety, and commences with axlebox design, where inadequate whitemetalling of the faces caused excessive wear, leading to increased sideplay, aggravating to the point of danger, the inadequate width tyres on the flangeless wheels.

A number of other mechanical and structural defects were investigated, and these will be dealt with in the summary of recommended modifications listed at the end of this account. Under VII, *General Operating Efficiency*, the Royal Commissioner becomes quite scathing in his remarks about Mills and his testing staff, who evidently had not the first idea as to how to go about testing a locomotive, nor as to what results might be expected. Evidently nobody bothered to calibrate the gauges and instruments employed, with the results that steam pressures appeared at the superheater side of the header when *drifting with regulator shut*, at which point one expects a slight vacuum, and despite the much longer steam piping, pressure drop to the rear steam chests was less than to the front. Some of these pressure drops recorded were as low as 3 to 4 lb/sq in, less than normally encountered with the best non-articulated locomotives, and with even the best Garratts elsewhere in the world, pressure drops of the order of 12–14 lb/sq in were normal, and calculable by standard flow-through-pipes formulae. Readings were 'corrected' to no good avail, and at one stage, a column of alleged temperatures turned out to be pressures from a different part of the locomotive, and relating to a different test! Under the circumstances, it is hardly surprising that the Commissioner commented: '. . . the data sent forward relating to all pressures and temperatures beyond the boiler is worthless and the conduct of the special tests was no credit to the Department'.

There is insufficient room here to summarise the whole report, but what will be of interest is the list of modifications recommended as necessary to make the engines really railworthy, and these follow below:

1. All locomotives to be checked for excessive wear between wheel hubs and axleboxes, and all those offending to be sent immediately to shops for rectification.
2. All leading coupled wheels to receive flanged tyres.
3. All intermediate unflanged wheels to have tyres with thin flanges.
4. Until flanged leading wheels were fitted, locomotives to be inspected once a fortnight.
5. Leading bogies to be entirely redesigned, and swing link suspension substituted.
6. Roller bearings to be applied to bogies.
7. Until leading flanges and new bogies have been applied, locomotives to be restricted to 25mph.
8. Side bearers to be provided at the main pivot centres, and the stresses to be recalculated for these bearers. (Some idea of the commissioner's opinion regarding the WAGR design department shows through in this instruction to calculate stresses in a load-carrying member, normal procedure in most drawing offices!)
9. Provision to be made for manual oiling of pivot centres.
10. Compensated spring rigging to be provided.
11. The tanks to be lowered as far as possible (about 1ft), and further baffles and stays provided.
12. Flexible firebox stays to be provided.
13. Condensate drains be fitted to steam brake cylinders, leakage of brake steam into the vacuum exhaust cones eliminated, better seating provided for the steam brake valve, plus additional leverage for the brake lever.
14. Reversing screws to be replaced by power reverse.
15. Better provision to be made for control and adjustment of the regulator.
16. Exhaust steam injector to be replaced by live steam injector.
17. Better gauge glasses giving more visibility, and better water cocks to be provided.
18. Coal chute to be altered.
19. Level taps to be provided in water tanks.
20. Bunker doors to be fitted.
21. Large cab ventilators to be provided.
22. Floor drains to be fitted.
23. Ventilated tucker boxes (food containers) to be supplied.
24. Means be provided to wet the coal.
25. Pivot centres to receive concertina covers to prevent ingress of dirt.
26. Shovelling plate to be fitted under coal chute.
27. Consideration be given to redesigning the long connecting-rod.
28. Improved spring-loaded release valves to be provided for cylinders.
29. More rigid levers to be provided for release valves and sand gear.
30. Better tank filler covers to be fitted, to prevent loss of water.
31. Sandboxes to be protected or altered in position to prevent ingress of water.
32. Speed recorders to be fitted.
33. Front ends to be regularly inspected to ensure that tubes and flues are kept clean. (This seems an amazing indictment of WAGR operation.)
34. All structures in districts using ASGs to be altered where necessary so that they conform to the fixed structure gauge, allowing adequare clearance from the loading gauge. (!)
35. That one locomotive be put under test for eighteen months to determine the best design of dome (to be fitted to domeless boilers), steam dryer, long steam pipes, and front end. All locomotives to be duly modified when optimum proportions have been determined.
36. Better trained staff and improved equipment to be provided to carry out the requirements of the previous item.
37. Costs of tests to be arbitrated, together with prices of locomotives sold to railways by the CLTB.

As it was, few of these recommendations were fully carried out. Of the using railways, both the Queensland and South Australian Government systems replaced them by Beyer Garratts as soon as these could be delivered, and in Tasmania and Western Australia they pottered on for a few more years, the last WAGR examples being withdrawn in 1957. The only railway to make a serious attempt to eradicate the defects was the little Emu Bay Railway in Tasmania, which already operated three Beyer Peacock 4-8-2+2-8-4 Garratts, and was presumably able to quietly apply existing know-how to the new locomotives. The remaining user, the Australian Portland Cement Co, in Victoria, probably had too limited an operation to severely try its ASG which, interestingly, was the sole survivor of the type in mainland Australia, and was at the time of writing in North Williamstown Museum.

The last survivor of the ill-fated Australian Standard Garratt awaits restoration at North Williamstown, Victoria, museum.

(*A. E. Durrant*)

Some background as to why the ASG swere built at all also appears in the Royal Commission's report. It seems that after it had been decided that the Garratt met the operating requirements of Queensland and Western Australia, the two states most closely involved, Beyer Peacock was invited to quote for supplying them. Two offers were made, one for £20,000 each to an unspecified design, and the other of £26,000 per locomotive to a 'Kenya' design, again unspecified, but probably to the old EC1 class. Apparently both these offers were considered too high, so Beyer Peacock then quoted for a set of drawings for the 'Kenya' type; the price of £5,000 plus a royalty of £3 per ton on every locomotive built was also considered too high, and Australia decided to go it alone. As has been seen, this proved a disastrous decision, especially as the saving in price came to about 10 per cent only on Beyer Peacock's lower offer. Tried and proven locomotives at £20,000 each are better than £18,000 defective hulks earning no revenue and awaiting extensive and expensive alterations. Looking through Garratt material, it is interesting to speculate on what designs Gorton offered. The Kenya EC1 was quite a lot bigger than allowed in Queensland, and would have needed extensive redesign. Incidentally, in Beyer Peacock's offers, any redesign needed would have been strictly under their supervision—evidently they were wary of the Garratt's good name being besmirched by 'wild colonial boys'. Looking back through details of the Garratts built by Beyer Peacock, only the Nigerian 900 class seemed light enough with some modification to run in Queensland, and this would have needed only a smaller boiler (perhaps that fitted to the ASG) and some redistribution of weight to have made it light enough. It is interesting to compare the dimensions of both the Nigerian and Kenya Garratts with the ASG and also with those eventually built new for Queensland:

Railway	*Nigeria*	*Kenya Uganda*	*Queensland*	*Commonwealth*
Cylinders (in)	$16\frac{1}{2} \times 23$	$16\frac{1}{2} \times 22$	$13\frac{3}{4} \times 26$	$14\frac{1}{4} \times 24$
Wheels	4ft 0in	3ft 7in	4ft 0in	4ft 0in
Pressure (lb/sq in)	180	170	200	200
Tractive effort at 85% (lb)	39,920	40,260	32,770	34,420
Grate area (sq ft)	35·8	43·6	39	35
Weight (tons)	119	125	137	119
Adhesion (tons)	76	79	77	68
Maximum axle load (tons)	9·5	10	9·65	8·5
Water (gallons)	3,800	4,250	3,800	4,200
Coal (tons)	5	6	6	6

From the above it may be seen that either of the two African light Garratts could have been pared down to the ASG axle loading without much difficulty, and in retrospect it was certainly a false economy not to use an existing design, especially in the urgency of wartime, when the odd couple of thousand pounds per locomotive was chicken feed compared with the millions spent on munitions.

COMMONWEALTH RAILWAYS

Class	*Type*	*Cylinders in × in*	*Wheels ft in*	*Heating Surface sq ft*	*Super-heating Surface sq ft*	*Grate Area sq ft*	*Weight tons*	*Pressure lb/sq in*	*Tractive Effort at 85% (lb)*
D	4-4-0	16 × 24	5 0	945	—	14·8	—	140	12,200
G	4-6-0	20 × 26	5 0	1600	—	27·0	58·2	160	23,600
GA	4-6-0	21 × 26	5 0	1370	252	27·0	58·2	160	26,000
K	2-8-0	22 × 26	4 3	1785	262	30·0	68·5	150	31,460
KA	2-8-0	22 × 26	4 3	1757	252	29·8	71·4	150	31,460
C	4-6-0	23 × 26	5 9	1955	650	30·5	87·5	180	30,500
CA	4-6-0	22 × 26	6 1	1929	400	34·6	78·1	190	27,840
CN	4-6-0	$20\frac{3}{8}$ × 26	5 3	1738	401	31·6	65·6	200	29,125
L	2-8-2	22 × 28	4 9	2595	619	47·0	96·5	200	40,400
*NF	2-6-0	12 × 20	3 3	544	—	9·8	19·6	120	7,530
*NFA	2-6-0	14 × 20	3 0	—	—	—	—	130	12,000
*NG	4-6-0	13 × 20	3 3	643	—	18·8	27·1	150	11,000
*NM	4-8-0	17 × 22	3 9	763	177	18·5	44·9	160	19,200
*ASG	4-8-2 +2-8-4	(4) $14\frac{1}{4}$ × 24	4 0	1698	315	35·0	115·9	200	34,520

*3ft 6in gauge.

Chapter 6

QUEENSLAND

The Queensland Government Railways were the first in Australia to adopt the 3ft 6in gauge. Not only this, but so far as the author is aware, the QGR was the first major system in the world built to this gauge. Three feet six inches is often known in Germany as *Kapspur* or Cape gauge, referring to the Cape of Good Hope in South Africa. However, it might be more appropriate to tie the term to Queensland's Cape York, for the first section of the system from Ipswich to Grandchester was opened in 1865, nine years before South Africa commissioned its first 'Cape gauge' power. Despite this seniority, development in Queensland lagged sadly behind other lines of this gauge, and all the other 3ft 6in lines in Australia overtook Queensland in the matter of powerful steam locomotives.

The first locomotives used were miniscule four-coupled efforts, the A10 class 2-4-0 by Avonside, supplied in 1865. It might be as well here to outline Queensland's strange and unique locomotive classification system in which the initial letter meant (in most cases) the number of coupled axles—not logically from A = 1, as in Japan, but A, B, and C meant 2, 3, and 4 coupled axles respectively, the letter being followed by a number to indicate the cylinder diameter in inches. D meant tank engines, leaving no room for possible ten-coupled types, and where more than one class existed with the same basic features, assorted prefixes were added, giving weird designations such as 6D13½ and BB18¼.

The four A10 2-4-0s were followed in 1866 by thirteen 0-4-2s of the same overall sizes, built by Neilson. One of these, sold out of service to a sugar mill, survived to complete a century of steam, and after taking part in the QGR's centenery celebrations, has now a place in the Redbank museum. Following on the A10 classes, Queensland experimented with the double Fairlie type, and three 0-6-6-0s were supplied in 1866 by James Cross of St. Helens, Lancashire, but trials of the first one were so unsatisfactory that all three were returned to England. An illustration and further details of these interesting articulated locomotives appear in R. A. S. Abbott's *The Fairlie Locomotive*. Another Fairlie, rejected by Norway, came out to Queensland in 1876, and seems to have been more successful, lasting until 1902 on coal haulage around the Ipswich area, and again Abbott's book gives full details.

The prototype 'Australian Standard' narrow gauge locomotive came out from Neilson in 1869, and was a very small 2-6-0, class B11, with the leading truck wheels placed behind the cylinders. Boilers had domes over the fireboxes, and very small four-wheeled tenders were attached. Only four were built, but they started a design standard which may be traced right through to the last steam engines built for Queensland, almost a century later, in 1958. Sixteen larger engines of the same basic design, class B12, came from Kitson in 1870, and this was the standard built also for South Africa and Western Australia, on the same gauge. Further batches of the A10 2-4-0, and the slightly enlarged A11 class, came out in the next few years.

In the late 1870s Baldwin's energetic representative won orders for five classes of locomotive. In 1878–79 the firm delivered three 4-4-0 class A12, two each of 'large' and small 2-8-0 classes C15 and C13, plus a couple of Moguls which took the same class number, B11, as their English predecessors. Evidently these first Baldwins were considered satisfactory, for in 1881–83 we find the same firm exporting two A10 4-4-0s, four C16 2-8-0s, and eighteen A12 4-4-0s, although after this the American firm dropped out of the picture altogether, either through their engines being unsatisfactory in the long run, or due to a less energetic agent being appointed.

Apart from odd tank engines of no particular interest, we next find Dübs building a C13 class in 1883 (note how entirely different classes received the same designation under Queensland's system) and the same year the firm also brought out a development of the earlier 2-6-0 types in the form of the B13 4-6-0. These looked very similar to the B12 class, with horizontal outside cylinders fed through overhead valves driven from inside motion via rocking shafts; all

Typical of early Queensland motive power, the Class B13 4-6-0 was a direct descendent of the original 2-6-0 types.

(*QGR*)

that really was done amounted to placing another pair of carrying wheels ahead of the cylinders, and connecting up as a bogie instead of a pony truck. The B13s were immediately successful, and some 112 were built, mostly by Dübs, plus others from Kitson, Phoenix, and by the railway's Ipswich works. As principal main line power, the B13 class ousted the B11 and B12 Moguls.

In 1884, Dübs built the most powerful engine to date, the 8D15 2-8-2T, the first noteworthy tank engine on the system. However, they did not last long in this form, probably being too heavy for the track, and were converted to 2-8-0 tender engines, class C15, in 1891. It was at this stage that Queensland began to be overtaken in the matter of 3ft 6in gauge motive power by Western Australia and across the Indian Ocean by South Africa, both of whose railways had based their earlier power on that proved successful in Queensland. At this stage, all three railways chose eight-coupled tank engines for the most severe duties, and it is interesting to compare the dimensions of the relevant classes:

Although the Queensland engines were the smallest of the three, they were also somewhat earlier, so the three types may be considered fairly comparable. This was the turning point, as a later comparison will show, and from this stage onwards, Queensland chose a path leading to the point where its 'Cape gauge' engines were much smaller than those of other Australian narrow gauge state systems, to say nothing of the nearest overseas neighbours such as New Zealand, Indonesia, and South Africa.

1889 saw the B13 class developed further as the B15 design, still with tiny driving wheels, but otherwise generally enlarged. Ninety-eight were built by English and Australian manufacturers, and they became the principal freight engine on the system, being followed soon after by two new 4-4-0 passenger designs, class A12 with 4ft 0in driving wheels, in 1890, and the larger A14 with 4ft 3in wheels, in 1894. Twenty-five and eight respectively were built to these two passenger designs, all by local builders for the first time in Queensland.

Railway	*Queensland*	*Western Australia*	*South Africa (Natal)*
Class	8D15	K	Dübs A
Type	2-8-2T	2-8-4T	4-8-2T
Date	1884	1893	1888
Cylinders (in)	15 × 20	17 × 21	17 × 21
Coupled wheels	3ft 0in	3ft 2in	3ft 3in
Tractive effort at 85% (lb)	13,750	18,900	18,500
Locomotive weight (tons)	37	53	47·2

Queensland Government Railways Class B15 was enlarged and improved from the earlier B13 4-6-0. The Class PB15 was the same design, with larger wheels for passenger work. *(QGR)*

1898 saw the opening of the Abt rack railway from Rockhampton southwest to Mount Morgan, the only rack railway on the Australian mainland, although the private Mount Lyell line operated in Tasmania. The first rack engines built, by Dübs in 1898, were 4D11½ class 0-4-2T to the same design as the Mount Lyell engines delivered two years earlier, and these were followed by six larger 0-6-0T, class 6D13½ built by Dübs and North British from 1900 to 1915. The rack line continued with its original motive power until 1952, when the 1 in 16 gradients were replaced by an easier adhesion-worked diversion.

1900 saw the introduction of the PB15 class, yet another 4-6-0, this time with driving wheels large enough for passenger work, the P standing for 'passenger'. Used on all classes of main line train, the PB15 became Queensland's most numerous class, 202 being built to the original design with inside Stephenson valve gear driving overhead slide valves through rockers, in the traditional Queensland manner. So successful were they that a start was made in 1900 to convert the B15 into a similarly versatile machine, enlarging the coupled wheels and raising boiler pressure to maintain tractive effort. With three-hundred engines between the two classes, the QGR was well provided with mixed-traffic engines. In 1901 it built twenty suburban 4-6-2Ts

The original locomotives of Queensland's Mount Morgan Abt rack line were the Class D11½ 0-4-2Ts built by Dübs. *(QGR)*

to supplement and replace the older four-coupled tank engines on the Brisbane services. Queensland was by this time becoming quite busy, and these six-coupled suburban tanks were contemporary with similar engines on the broad and standard gauge systems.

Apart from the early 2-8-0s with shirt-button size driving wheels, Queensland still had no eight-coupled power for main line work, but in 1903 produced the C16 4-8-0, a substantial advance over earlier six-coupled power for freight work, whilst with leading bogie and larger wheels, the design was fast enough for Queensland's heavier passenger duties. The years 1902–04 saw a number of notable 3ft 6in gauge 4-8-0 designs placed in service; in Australia the WAGR introduced their F class in 1902, whilst the South Australian T and QGR C16 followed in 1903. Comparing with South Africa, the Cape Seventh class was of 1892 vintage, small engines, although comparable in size with the WAGR F and South Australian T classes. More important were the Cape Eighth class of 1902 and the Natal B built 1904, both of which began to show the way as to what could be done within the confines of 'Cape gauge'. The Natal engine was far more powerful than any contemporary broad or standard gauge locomotive in Australia, let alone the narrow gauge types, and from this point onwards, no Australian narrow gauge engine could be compared with the best in South Africa. The various 4-8-0 types mentioned in the above paragraph are compared dimensionally in the table below:

Railway	*Cape G*	*WAGR*	*South Australia*	*QGR*	*Cape G*	*Natal G*
Date	1892	1902	1903	1903	1902	1904
Class	7th	F	T	C16	8th	B
Cylinders (in)	17 × 23	17 × 23	$16\frac{1}{2}$ × 22	16 × 22	$18\frac{1}{2}$ × 24	$20\frac{1}{2}$ × 24
Wheels	3ft $6\frac{3}{4}$in	3ft $6\frac{1}{2}$in	3ft 7in	3ft 9in	4ft 0in	3ft $9\frac{1}{2}$in
Pressure (lb/sq in)	160	160	185	175	180	190
Tractive effort at 85% (lb)	21,150	21,250	21,150	17,500	26,200	35,450
Grate area (sq ft)	17·5	19	17·3	18·5	21·35	34

In one respect the QGR C16 class was of modern design, and that was the use of outside Walschaerts valve gear, although of the engines in the above table, both WAGR F and Natal B class were similarly equipped.

Also a Glasgow product were the enlarged Mount Morgan Abt rack locomotives, the Class D13$\frac{1}{2}$ 0-6-0Ts.

(*QGR*)

An interesting half-way stage between 4-6-0 and Pacific was the Class B16½ 2-6-2. Unusual features included Southern valve gear and bridle frame around the firebox. *(QGR)*

Below: Featherweight and versatile, the QGR Class C17 4-8-0s were built over a long period. No 974 is shown here near Gympie with a tour train at Easter 1972. *(A. E. Durrant)*

Despite their small dimensions, the C16s were excellent locomotives, and set a new design standard comprising outside cylinders with overhead slide valves actuated by Walschaerts gear, fed with saturated steam from a round-topped boiler. Within these standard fell the twenty-one enlarged passenger 4-6-0 class B17 of 1911, but the standard was soon improved to substitute

piston for slide valves. In 1914 three heavy passenger 4-8-0s of class C18 were built to the new standard, these being Queensland's first piston valve design.

Increasing demands for suburban engines caused the 6D16 4-6-2Ts to be converted to 4-6-4Ts in 1905, and in 1917 a PB15 was also rebuilt to the same wheel arrangement and classed B15D. The following year saw Ipswich outshopping a new express engine, class B16½, a 2-6-2 of neat appearance spoilt only by Southern valve gear. This was the second and last 2-6-2 tender engine built for Australia, and the first Queensland design to have a wide firebox. Around the firebox was a type of bridle frame, not dissimilar in principle to that used by Beatty and Hendrie in South Africa, although on the QGR engine it was of built-up rather than cast construction, and extended forward of the trailing coupled wheels. A superheater was included for the first time, steam being conveyed from the header to steamchest by adequate external pipes. It was altogether a modern little design, spoilt only by the valve gear, possibly the cause of the locomotives' non-multiplication, although easily replaceable by Walschaerts motion.

In 1920 the C17 4-8-0 came out as a superheated piston valve version of the earlier C16, and so useful did these lightweight engines prove to be that 227 were built altogether, the last as late as 1953. Denizens of other Australian states tended to laugh at the tiny Queensland locomotives, and called them 'clockwork engines' but they did have remarkable powers of speed and haulage in relation to their size.

The Class C18 4-8-0s were the first 'heavy' locomotives in Queensland. They were with piston valves, but unsuperheated; later they were superheated and brought into line with the C19s.
(*QGR*)

The author's experience with QGR steam was all too brief, but the performances noted were certainly better than the size and appearances of the locomotives might suggest. While on the subject of QGR 4-8-0 it is convenient to chronicle the two larger varieties of classes C18 and C19, introduced in 1918 and 1922 respectively. Only three of the C18 class were built, all in the railway's Ipswich workshops, and they were designed as a heavy passenger engine particularly for the Sydney mail connection via Wallangarra, a steeply-graded route over the Toowoomba mountains, which served as the main line until the coastal standard gauge route was completed much later. Although they had piston valves and Walschaerts gear, the C18s were unsuperheated, and remained in this form until 1934, when superheaters and larger cylinders were fitted to bring them into line with the later C19 class. The rebuilds were classed CC19. The superheated C19 series closely resembled their predecessors in appearance and were for similar duties, but with twenty-six engines in the class more universally employed on heavy freights as well as the Sydney mails. 'Heavy' is, of course, a relative term, for while the C19s were the largest and most powerful non-articulated steam engines to run in Queensland, their axle load was a modest 10 tons, the equivalent of branch line power elsewhere in the world. It is amusing to read in *The Redbank Museum* booklet the statement that the C19 class was 'among the largest engines on the 3 feet 6 inches gauge'. Although it may seem unkind to deflate the Queenslander's doubtless sincere patriotism, it must be pointed out that by the time the initial ten C19 engines were delivered, South Africa had in service nearly *three hundred* conventional 4-8-2s of approximately *double* the C19's capacity, plus the even more powerful MH Mallet and GA Garratt classes! The C19s were

in fact more in line with the Cape Government Eighth classes of twenty years earlier. In case these bald statements should agitate loyal Queenslanders, it is only right to quote chapter and verse in suitable tabular form below.

Railway	*QGR*	*SAR*	*SAR*	*SAR*	*SAR*
Class	C19	12	12A	14	14C
Type	4-8-0	4-8-2	4-8-2	4-8-2	4-8-2
Number in service (1924)	10	76	15	100	73
Tractive effort at 85% (lb)	23,525	41,680	47,420	42,340	42,340
Grate area (sq ft)	21·4	40	41	37	37

The years 1924–26 saw the introduction of three new classes, the first of which was the D17 4-6-4T suburban engine, with larger driving wheels than most main line classes. Some thirty were built, and they lasted until the end of suburban steam in Queensland. A like number of light branch engines were built by Walkers in 1925 and although classed PB15 were an advance on the older engines carrying the same designation, for the new design had outside Walschaerts gear driving overhead slide valves.

The final new type of this period was also the most important, being Queensland's first Pacific, the B18¼ class. Of the Australian states which adopted 3ft 6in as its major gauge, Queensland was the last to introduce the Pacific type for passenger work; the engines were, predictably, the smallest and least powerful. This is not to say they were inferior, and the little B18¼s have been known to speed along at about 70mph despite their small driving wheels. Cylinders, motion and framing were generally to the standards of the 4-8-0 classes, but the 4-6-2s had wide fireboxes, for the first time on the QGR other than the odd B16½ class. A total of eighty-three were built, all in Queensland by Ipswich works and by Walkers, and as it transpired it was the last completely new class to be designed and built in Queensland. From their appearance until the 1939 war, only rebuilds were added to the list of classes, the CC19 4-8-0 of 1934 rebuilt from earlier C18 engines, the B16D 4-6-0 converted from 6D16 4-6-4T, and another tank-to-tender conversion for shunting, from 6D13½ 0-6-0T to B13½ 0-6-0.

The Pacific war of the 1940s imposed a great strain on Queensland's motive power, for traffic almost doubled within a very short period. The QGR engines were extremely hard pressed to

Below: Pictured here, the Class B18¼ and the later BB18¼ were speedy locomotives on the QGR 3ft 6in lines. Painted in GWR green, a B18¼ heads an Easter 1972 tour north of Brisbane. (*A. E. Durrant*)

Crane tank locomotives were once a feature of the Australian scene. In NSW and Victoria they have survived main line steam traction. This early Beyer Peacock example once worked in Queensland.

(*QGR*)

cope, and it was obvious that something much more powerful was needed. Accordingly, the ASG Australian Standard Garratt was designed and built, with Queensland particularly in view, and as this controversial type was dealt with in greater detail in chapter 5, we need only chronicle here the fact that twenty-three actually went to Queensland, and were only used for a few short years.

As a stop-gap measure, twenty American 'MacArthur' 2-8-2s were also sent to Queensland in 1943, and these proved very useful, although no more powerful than the larger QGR 4-8-0 types. To avoid confusion with the existing C16 class, they were known as AC16s, and had slightly less tractive effort but substantially more boiler capacity than the C19s. No other Australian railway used this war engine, although all round to the north, in India, Burma, Malaya, and Thailand the same design built to metre gauge was found by the hundred.

After the war, twelve enlarged suburban 4-6-4Ts, class DD17, came out of Ipswich in 1948, and these formed the last class of steam suburban engines built new for Australia. More important, in 1950, a new Garratt class of Beyer Peacock design was introduced, despite the poor showing made by the wartime ASG type. It must have taken some courage to reintroduce the Garratt so soon after the previous locomotives of the species had been tried and found wanting, but if nothing else the ASGs had shown how the Garratt could haul far heavier trains than conventional engines of the same axle load. The new Beyer Garratts were built with what in Queensland amounted to express passenger driving wheels, and they were in fact used on express work as well as on the heaviest coal haulage. Without a doubt, the later QGR Garratts were a success, and lasted until the end of steam. The last new class to appear, the BB18$\frac{1}{4}$, was a development of the earlier B18$\frac{1}{4}$ 4-6-2s, but substantially improved with longer-travel piston valves, roller bearings on all axles, and SCOA-P wheel centres. All basic dimensions were as on their predecessors. The first batch came from Vulcan foundry in 1951, and numbered thirty-five engines, after which Walkers commenced deliveries of twenty engines of which the last, 1089, delivered in March 1958, had the melancholy distinction of being the last new steam locomotive built for an Australian railway. A new CC19 class 2-8-2 was designed and eighteen ordered but later cancelled.

In the later post-war years Queensland embarked on an attractive programme of painting locomotives in assorted bright colours, according to class. Pacifics appeared in a rather Great Western shade of green, C17 4-8-0s in reddish brown, Garratts in maroon, and 4-6-4Ts in a light metallic blue, although many engines lasted in black until the end of steam in 1969.

In addition to its main 3ft 6in gauge network, the QGR also ran a 2ft 0in line, the Innisfail Tramway with about twenty little engines, mostly 0-6-0T and 0-6-2 by Fowler, a Krauss 0-4-0WT, and four Hunslet 4-6-0T of World War I vintage.

QUEENSLAND: TYPICAL DIMENSIONS OF SELECTED CLASSES

Class	*Type*	*Cylinders in × in*	*Wheels ft in*	*Heating Surface sq ft*	*Super-heating Surface sq ft*	*Grate Area sq ft*	*Weight tons*	*Pressure lb/sq in*	*Tractive Effort at 85% (lb)*
A10	0-4-2	10 × 18	3 0	396	—	7·1	13·6	120	5,100
B12	2-6-0	12 × 18	3 3	538	—	10·0	17·7	140	7,900
B13	4-6-0	13 × 20	3 3	643	—	18·8	27·1	140	10,300
B15	4-6-0	15 × 20	3 0	808	—	12·8	31·6	140	14,850
PB15	4-6-0	15 × 20	4 0	872	—	13·1	34·0	160	12,750
B15 Conv.	4-6-0	15 × 20	3 9	808	—	12·8	31·6	160	13,600
B16½	2-6-2	16½ × 22	4 3	1363	303	25·7	53·0	175	17,400
B17	4-6-0	17 × 23	4 0	1268	—	19·0	—	175	20,600
B18¼	4-6-2	18¼ × 24	4 3	1520	334	25·3	56·4	170	22,650
BB18¼	4-6-2	18¼ × 24	4 3	1524	334	25·0	58·0	170	22,650
C16	4-8-0	16 × 22	3 9	1029	—	18·5	44·0	175	18,600
C17	4-8-0	17 × 22	3 9	763	177	18·5	44·9	175	21,000
C18	4-8-0	18 × 23	4 0	1340	—	21·4	52·7	175	23,100
C19	4-8-0	19 × 23	4 0	1043	225	21·4	54·3	160	23,500
CC19	4-8-0	19 × 23	4 0	1014	192	21·4	52·9	160	23,500
AC16	2-8-2	16 × 24	4 0	1371	374	27·7	53·5	185	20,100
6D13½	0-6-0T	13½ × 20	3 0	525	—	10·0	30·6	140	12,050
D17	4-6-4T	17 × 22	4 3	979	145	16·6	58·3	170	18,010
DD17	4-6-4T	17 × 24	4 3	1064	150	18·5	62·0	180	20,810
Beyer Garratt	4-8-2 +2-8-4	(4) 13¾ × 26	4 3	1668	453	39·0	136·8	200	32,770

CHAPTER 7

WESTERN AUSTRALIA

With its tiny population, it is rather remarkable that money was found to commence railway operations in Western Australia as early as actually happened. The first line built was not a public railway at all, but a privately-owned and operated timber hauler. However, it did set the state gauge at 3ft 6in, and although its only locomotive never came into Government railway ownership, part of the route now forms a section of the Nannup branch line. Opened in 1871, the Western Australian Timber Company's line ran twelve miles from Yokonup to the coast at Lockeville, near Busselton, crossing the present Bunbury–Busselton line near Wonnerup. The sole steam locomotive was an extraordinary long-wheelbase 0-4-0 built at Ballarat, Victoria, and shipped to Western Australia in 1871. The decaying hulk of this interesting pioneer now rots away in the park at Busselton.

The first state-owned public railway was that from Geraldton to Northampton, about three-hundred miles north of Perth. Mineral traffic, particularly lead, was supposed to justify the line's existence but the venture was not a success. Motive power started with two Kitson 2-6-0s of the same type as built for Queensland and South Africa; when the line closed in 1896, these became class M on the WAGR. There were also two Fairlie engines of the unique 2-4-4-2T type, built by Avonside in 1879; failing to satisfy, they were transferred to the main system in 1885 and 1892. Technical details are to be found in Rowland Abbott's *The Fairlie Locomotive*, to which can be added the fate of the first engine, sold to a timber line in 1893 and afterwards lost trace of, while the second engine upon transfer south was rebuilt to a 2-4-2T and also eventually sold. As Fairlies

The first locomotive in Western Australia was *Ballarat*, built in that town for a private timber railway, part of which now forms a section of the WAGR Nannup branch. The locomotive is shown here in Busselton Park. (*A. E. Durrant*)

KATIE

G233

Top left: WAGR No 1 *Katie*, beautifully restored, at Bassendean museum. (*A. E. Durrant*)

Centre left: Cast in the Beyer Peacock mould, narrow gauge 2-6-0 No G233 was an enlargement from the earlier Class A, and built in Australia by Martins. No 233 is still steamed for the 'Leschenault Lady' vintage train run by the Bunbury Tourist Bureau, and is seen here in 1970 at Brunswick Junction. (*A. E. Durrant*)

Bottom left: The 4-6-0 'Bogie G' was of the same dimensions as the 2-6-0 version. No 71 was built new to the WAGR design for logging work, and is shown here at Millar's timber yard in 1971. Lasting until 1973, it was the last steam locomotive in regular service in Western Australia. (*A. E. Durrant*)

Above: Class H 0-6-0T No 18, painted-up at Bunbury depot, awaiting official preservation. (*A. E. Durrant*)

they were class E, and the rebuilt engine class F.

For the first WAGR main line, Fremantle–Perth–Guildford, two 0-6-0ST were built by Robert Stephenson & Co., with the peculiar flat-topped saddle tank favoured by that firm for shunting engines. Although used on the short 'main line', they were really standard English industrial shunting engines reduced to narrow gauge. Number 1 *Katie* had a long and varied career, being converted to a tank-tender engine at one stage, sold to a timber line, and eventually returned and restored at the Bassendean railway museum, where it overlooks its original operational route.

As the line, known as the Eastern Railway, spread inland towards York, tender engines were needed, and the classic Beyer Peacock small 2-6-0 was introduced and classed A. These survived for a long time as branch engines after displacement from the main lines. For heavier work over short distances, Kitson built eleven B class 4-6-0Ts in 1884–98, and other miscellaneous tank engines of the early period were 0-4-0ST (D), 0-6-0T (H), and 0-6-2T (L). In 1891 three 0-6-4T single Fairlies were bought from New Zealand, becoming class I, although lasting only until 1900. Much more useful were the three J class Kitston 4-6-0 tender engines, having many features similar to the B class, but with larger wheels and cylinders. Later they were rebuilt with Q class boilers, and reclassed JA.

So far, the engines employed had been an odd collection of unstandardised classes which must have been quite a headache to maintain, especially for somewhere so remote as Western Australia. The first really important class numerically was the G class 2-6-0, similar to but larger than the A class, and a Beyer Peacock design the same as South Australia's Y class. The G class seems to have suited exactly conditions in Western Australia, for we find that from 1892 to 1898 over fifty were supplied, plus about twenty more of a 4-6-0 version, still called the G class, although one would have expected some sort of distinction such as GA. The 1890 decade saw tremendous expansion of the WAGR system, and the first eight-coupled locomotives were some twenty-four K class 2-8-4Ts built by Neilson in this period. Six more were ordered from the same builders, who by the time the order was executed had become part of the North British Locomotive Company. These final examples of the class, built in 1902, were diverted to South Africa for use in

The passenger version of the Class G was the Class R 4-4-0. No 174 is preserved in blue livery outside Midland Works. For a while they ran as the only Atlantic class (4-4-2) on an Australian state railway.
(*A. E. Durrant*)

the Anglo-Boer war, where eventually they became Central South African Railways class C. Although popular in Western Australia, and even used on suburban passenger trains, especially when the Claremont show was on, South Africa found them too slow, and sold all to mines and similar users after less than ten years' service.

1896 saw no less than five new classes introduced. Two 4-4-0 passenger engines of class P came from Martins in South Australia, the same as the South Australian Railway's Z class. In 1912 both were sold to the Midland Railway of Western Australia. Neilson supplied the first of thirty-two N class, delightfully neat 4-4-4Ts for suburban work and, using the same sized boiler and cylinders, the curious O class 2-8-0 tank+tender design, eventually to number forty-six locomotives. The O class locomotives were more powerful than the K class tanks and had the advantage of greater range due to tenders being coupled on. The exact reason for adopting the tender+tank arrangement is now obscure, but presumably the N class boiler proved too small to provide adequate rear-end adhesion, and the tank acted mainly as deadweight. A larger boiler would have been more advantageous, but such was never fitted, although a few were superheated and reclassified OS. In 1908, after the advent of larger main line freight power, fourteen O class were rebuilt to N class 4-4-4T.

1896 saw also a new 4-4-0 express engine, the R class, similar to but larger than the earlier P. Although cast in the classic Beyer Peacock mould, all twenty-four were Dübs-built in 1896 and 1898. Between 1909 and 1928, fourteen were rebuilt with trailing pony trucks, making them the only Atlantics to run on a main line railway in Australia. By this time they had been displaced from the major passenger duties, and the conversion was probably to improve riding when running tender-first on local workings. The final class to appear in 1896 was a Hawthorn Leslie 4-6-2T whose introduction came about by accident, and formed one of the links between the 3ft 6in gauge railways of Australia and Southern Africa. From Lourenço Marques the Caminhos de Ferro de Norte had been formed to open up the Limpopo valley, and to connect inland to Rhodesia and the Eastern Transvaal. For some reason the venture failed, and of the ten 4-6-2Ts built, four stayed in Mocambique to become C. F. Lorenço Marques Nos 15–18, while the other six were snapped up by power-hungry Western Australia, to become their class Q. Evidently the latter's climate was more suited to these engines than that of south east Africa for the original four were withdrawn in 1912–14, whereas the Australian batch lasted longer, four of them being rebuilt as class Qa 4-6-4T in 1905–09. Valve gear was of the Joy type, unusual on outside-cylinder engines, and actuated overhead

Suburban tank development is shown here in Class N 4-4-4T No 200, an example of turn-of-century passenger motive power. (*WAGR*)

slide valves. Main duties of these interesting engines were suburban passenger and freight turns.

The final important act of 1896 was the purchase of the Great Southern Railway, built on the land grant principle, south to Albany. The Great Southern's main line power comprised ten 4-4-0s by Beyer Peacock and Kitson, generally similar to the existing P and R classes except for dimensional differences, and these became WAGR class T. For shunting there were two of Kitson's strange little tank engines with patent valve gear based on Walschaerts, but having the primary motion to the expansion link provided by a bell crank extension driven directly from the main crankpin. There seems little to commend this gear—it *did* eliminate return cranks, but it occupied an enormous amount of space, while valve events were affected by the vertical movement of axleboxes within hornblocks. These two engines became class S; they were sold out of service, ending up in industrial use in New South Wales.

We now enter a controversial period, during which the world's first 'Pacific' type engines were placed in service, and where Western Australia stakes a claim against the rivals of America and New Zealand. It all depends, one supposes, on how one defines a 'Pacific'. To the author, any 4-6-2 tender engine is a 'Pacific', but there are those who aver that the trailing truck must support a wide firebox, in which case Collin's famous engines in northern France were not 'Pacifics' at all! What appears to be the very first 'Pacific' was a strange machine built by the Lehigh Valley Railroad, USA, in 1886. Incorporating Strong's patents, the wide firebox contained twin corrugated furnace flues, and was fair and square over the trailing truck. Cylinders were fed and exhausted through complex gridiron valves, and although these features were not a success, it is hard to deny that this was the first 'Pacific'. New Zealanders like to claim their Q class, Baldwin-built in 1901, as the first 'Pacifics', but the Karri & Jarrah forestry line in Western Australia had a 3ft 6in gauge 4-6-2 by Baldwin, built in 1898. These were followed in 1901 by the WAGR Ec class, twenty Vauclain compounds with narrow fireboxes, also by Baldwin. Whatever the rival claims by the using railways, there is no doubt that all the earliest 4-6-2s and/or 'Pacifics' were American-built, and the author's vote goes to the Strong engine as being the world's first. WAGR's Ec class suffered the usual Vauclain compound troubles, but soldiered on for over twenty years until in 1924–28 they were given a new lease of life by a thorough rebuilding. Boilers were superheated, and new simple-expansion piston valve cylinders replaced the originals. At the same time, the original Stephenson valve gear was replaced by Walschaerts, and they became quite respectable locomotives. The year after the Vauclains were built, Baldwin

ES 347

FS 460

supplied twelve conventional 4-6-0s of modern design with piston valves and Walschaerts gear, plainly derived from New Zealand practice, and similar to the various NZ 'U' classes, although the valve detail was more in line with the NZ 'Q' class. From 1908 to 1920 the C class were rebuilt with trailing pony trucks and classed Ca, and ten new Ca were built in the railway's Midland workshops near Perth in 1915. Finally, some were superheated to class Cs, and four of these, sold to a timber concern, lasted until the late 1960s.

These Baldwin designs were ordered by T. F. Rotheram, the new Chief Mechanical Engineer, appointed in 1900 straight from the corresponding position in New Zealand. The Baldwins were presumably a stop gap to fill motive power needs while Rotheram's own designs were under construction in Britain. Two classes were built, sixty-five 4-6-2s of class E and fifty-seven 4-8-0s of class F, each class being highly standardised with the other and differing only in the wheel size and

Top: The first modern locomotives in Western Australia were the E and F classes. Here a pair of Class Es head a freight through Midland, under a typical WAGR somersault signal. (*WAGR*)

Bottom: The last Class F 4-8-0 in service was No 460, seen here in 1971 at Collie on a trip working from the local sawmill. (*A. E. Durrant*)

arrangement. Plate frames and narrow fireboxes were conventional British features, but piston valves and Walschaerts gear were fitted. Rotherham imported features from New Zealand; the F class was clearly a development of the NZ B class introduced under Rotheram's chieftainship in 1896. Here again, we find New Zealanders claiming priority in combining the use of piston valves and Walschaerts gear in these 1896 machines, but once again it is a false claim as the old Chemins de Fer de l'Ouest, in France, was using Walschaerts, piston-valved engines as early as 1883.

With the E and F classes revolutionising traction in Western Australia, it is not surprising to find a substantial period when few new locomotives were supplied, although construction of the E and F classes was in two concentrated periods, 1902–03 and 1912–13. Between these batch dates it is surprising to find Midland works taking the retrograde step of building ten more obsolescent 2-8-0 tank + tender engines. These had slightly larger wheels than the O class, and were classed Oa; possibly they were an amalgam of spare components, assembled to provide some cheap branch line power.

After the last E and F classes were ordered, we find the WAGR concentrating on modern power for secondary services, and in 1912 there appeared the first new designs for suburban and light outback branch services. The suburban engines, 4-6-4T class D, were North British built to a neat design with Walschaerts gear and piston valves, features included in all subsequent WAGR designs, and requiring no further comment. The twenty D class were a substantial advance on previous suburban engines, and most were subsequently superheated and classed Ds. Similarly, the superheated Rotheram main line engines became Es and Fs, few remaining in original saturated condition.

The branch line engine was a completely new departure for Western Australia, and was a bold step, introducing the Garratt type. At that time, the Garratt was very much unknown and untried, with two compounds supplied by Beyer Peacock to Tasmania, and a solitary simple for the Darjeeling–Himalaya line in India, all three being 0-4-0+0-4-0 engines on 2ft 0in gauge. To order, at one go, six substantially larger engines of this new type points at either courage or foolhardiness on the part of the WAGR authorities, and good salesmanship on the part of Beyer Peacock. As it was, the M class 2-6-0+0-6-2 Garratts was an undoubted success. In many ways, the M class was the equivalent of two old A class 2-6-0 back-to-back, with a large boiler slung between, wheel diameter and cylinder dimensions being almost identical, so that the WAGR knew what it ought to get in the way of haulage capacity. These early Garratts were not superheated, but seven superheated Ms developments came out from Gorton in 1913–14. Jumping the intervening years, a further ten branch line Garratts, class Msa, were built in Midland works in 1930, to a slightly enlarged design, and these light Garratts were the mainstay of the Southwest's heavily-graded rural branches until displaced by W class 4-8-2s in the 1950s.

The final stage of WAGR steam development when superheated wide-firebox boilers were applied to conventional locomotives, was heralded by the P class 4-6-2 from North British in 1924. An improvement on the previous narrow-firebox Pacifics, their tractive effort was greater than the F class 4-8-0s, making them suitable for the heaviest freight traffic, as well as the fastest passenger trains. Ten were built in Glasgow, and

Above: The last Class S 4-8-2 in service was No 549 *Greenmount*, pictured in 1971 at Williams on the early morning freight from Narrogin to Collie. Note the bifurcated driving wheel spokes. (*A. E. Durrant*)

Top right: The last steam suburban locomotives were the Class Dd 4-6-4Ts, improved from the earlier Class Dm. Dd592 is shown here in the Perth suburbs in 1971. (*A. E. Durrant*)

in 1927–29 fifteen more came out from Midland works. In 1938–39, another ten to an improved and more powerful variation, class Pr, were built at Midland, followed soon after by eight rebuilds from P to Pr.

In 1943 a new concept came to Western Australia in the shape of the S class of similar size, power, and features as the Tasmanian Q class of 1922, also of the 4-8-2 type. The S class, known as *Sammies*, did have some unusual features, among these being domeless boilers, and a skyline casing over the boiler mountings. Ten were built by Midland between 1943 and 1947, and they handled heavy freight and passenger trains, at first on the Eastern districts. One of the *Sammies* achieved the sad distinction in 1970 of being the last steam locomotive in Western Australia to receive a heavy overhaul, and the class was the last in the state to remain intact. By 1971 all were stationed at Collie, being employed on trains to and from Wagin, Narrogin, and Bunbury. Contemporary with the S class were the Australian Standard Garratts, designed at Midland by the WAGR CME, and of which more is recorded in chapter 5. It is sufficient to say here that twenty-five ran in Western Australia, some attaining only six years' service.

After the 1939–45 war two classes of suburban tank engine first appeared, each of the 4-6-4T type. The eight engines of class Dm were rebuilds from E class 4-6-2, and used many components of these outdated main line engines, including outside admission cylinders with Z-chaped steam ports. The subsequent Dd class, built new, was similar in most respects to the rebuilds, but had a redesigned front end with straight-ported cylinders and inside-admission piston valves. Ten of

Right: The long line of WAGR Pacifics ended with the post-war Classes Pm and Pmr. Only three remained in regular service in 1970, when Class Pm No 701 was photographed with a Narrogin–Collie freight. (*A. E. Durrant*)

592

PM
701

Modern, lightweight, and versatile, the WAGR Class W 4-8-2s were in their prime when overtaken by diesels. A pair are shown here in 1971 between Donnybrook and Boyanup, with a freight from Bridgetown. (*A. E. Durrant*)

these latter were built, all at Midland.

Three new classes of 4-6-2 entered service during 1946–50, and the first of these added yet another link with Africa, for they were to the standard Pacific design used on the Sudan Railways. Most of these engines were eventually employed in Sudan itself, but twenty remained surplus at the war's end. Of these, six were bought by the Trans Zambesia Railway linking Moçambique with Nyasaland, and the remaining fourteen were bought by the WAGR, which designated them U class. Although not so powerful as the native P and Pr classes, the Us were swift and free-running, and soon became the system's principal express engine. Duties included hauling the 'Australind' train from Perth to Bunbury, the fastest narrow gauge train in Australia, taking $3\frac{1}{4}$ hours for 115 miles. Twenty-five years later, diesel traction had pared five minutes from this timing, less than the time saved by the elimination of the steam locomotives' water stops.

Soon after the U class, in 1950, North British delivered thirty-five new Pacifics based on the Pr class and having the same nominal dimensions. Unfortunately, one of the 'improvements' consisted of Cartazzi slides for the trailing axle, controlling side play by forces on inclined planes. Invented in Italy, the Cartazzi principle never found much favour elsewhere except perhaps on the LNER, whose Pacific and Prairie classes used the Italian slides. However, the horrible grinding and squealing noises accompanying a Gresley Pacific while negotiating a sharp crossover were indicative of flanges binding on the rails. On the V2 class, with only a swing link leading pony truck, conditions became dangerous after the war, when track maintenance standards had declined, and a number of derailments occurred. With such data to hand, it is surprising that such a device was adopted in Western Australia, whose poorly-laid and badly-maintained track needed the most flexible and road-holding chassis that could be provided. Not surprisingly, the Pm class locomotives proved too stiff for high speeds, and what should have been Western Australia's top passenger power became relegated to freight work. The series was divided into two sub-classes, Pm with roller bearings on carrying axles only, and Pmr with the same feature on coupled axles as well. All were intact in 1968, but by the end of 1970 were down to three examples used from Collie and Bunbury depots.

The subsequent few years saw Western Australia's last steam designs appear. First were sixty engines of the excellent W class light 4-8-2, whose axle load was just under ten tons. Beyer Peacock evidently had very much a free hand in the design, for the frame, cylinders, wheels and motion were virtually identical with that firm's light Garratt, as built for East Africa (60 class), and South Australia (400 class), except rear of the coupled wheels, where essential differences existed due to one being a Garratt and the other a conventional design. Large boilers with combustion chambers and thermic syphons provided plenty of steam, even with the low grade Collie

By 1971 steam had officially disappeared from the Perth–Bunbury line, although Class V 2-8-2s sneaked up to Perth on night freights. An unbalanced working over Easter produced, for the last time, two Vs returning together to Bunbury. (*A. E. Durrant*)

coal used as fuel, and these were probably the best motive power investment made in Western Australia, light enough for branches, as fast as any of the 'Pacific' types, due to excellent cylinder and valve design, and lacking only the tractive capacity for the heaviest freights. The W class might have been Western Australia's last steam class, for in 1954 the X class diesels appeared. However, the sensible decision was made to try out the diesels at first in the northern areas remote from coal supplies, and suffering water shortages. It was as well this was done, for the diesels suffered so many troubles that no more of any type were placed in service until 1960.

Meanwhile, the last and largest of Western Australia's steam classes, the V 2-8-2, were delivered from Robert Stephenson & Hawthorn Ltd, under sub-contract, the original order having been placed with Beyer Peacock. There had been schemes for three classes of the V series, a Vp passenger pacific, Vf freight engine, and Vg 2-8-2+2-8-2 Garratt, but unfortunately only the freight type appeared. Twenty-four engines were built in 1954–55, placed in service 1955–56. For ten years the V class locomotives were kings of the rails in Western Australia, efficiently handling the heaviest freights whilst burning low-grade Collie coal.

This was not good enough for the diesel manufacturers and oil companies, to whom in the late 1960s, Western Australia represented a failure. Between 1957 and 1967 only twenty-nine new diesels had been commissioned, and most of the railway was steam powered, with just about half of the steam engines of post-war vintage. A new steam roundhouse had been built at Collie, centre for the coal mining industry, and a start made to install CTC on the heavily-graded single track between Collie and Brunswick Junction, key bottleneck in the route for coal between the mines and Perth's power stations. Evidently the railway was not paying sufficient attention to the greasy god of mineral oil, and had to be taught a lesson, for blasphemous forecasts had foretold a continued reliance on steam in the southwest for a good number of years. Clearly political intervention was needed. The oil racket opened three fronts in Western Australia—power generation, buses, and of course railways. In the city transport field, repairs to diesel buses were charged to the trolleybus account, making the latter 'uneconomical', and 'justifying' a change from clean quiet electric power to noisy, polluting diesel buses. In the electrical generation field, the power stations in and around Perth were changed from coal to oil burning, and some fuss *was* made about this. After all, why import oil from overseas when there is perfectly useable local coal to do the same job? Somebody actually had the temerity to stand up and ask for comparative figures on the cost of electrical generation, using imported oil versus local coal but an answer to this leading question was 'not in the public interest', a political euphemism meaning that it was very much in the public interest. After a change in local government the figures were

The setting sun glints on Class S 542 and Class V 1209 when they double-head a Brunswick–Collie freight in 1971, Western Australia's last half-year of steam traction. (*A. E. Durrant*)

revealed, but by this time it was too late, and the damage had been done.

On the railway side the first blow came with the change to oil-generated electrical energy, for suddenly the major item of traffic on the Collie–Perth route disappeared, rendering many steam locomotives redundant. After that there was the wheat quota reduction, for in Australia, despite the fact that neighbouring Asia houses millions of starving humans desperate for any form of sustenance, farmers are only allowed to grow a limited quota of wheat for political and economic reasons. Western Australia, where wheat was moved by steam power, was clearly in a disadvantageous position in obtaining a wheat quota. When the wheat quota was severely reduced in Western Australia, again throwing more steam locomotives out of employment, it was simply a matter of attrition, and the eighty-five steam locomotives active in June 1970 were brought down to exactly zero twelve months later, although a handful were stored 'serviceable' at Collie.

Thus ends the sad tale of WAGR steam, whose final and finest were massacred prematurely to satisfy the vanities of American oil interests. Western Australia had the best 3ft 6in gauge steam in the sub-continent, and in latter years clad them in an attractive larch green livery. The money wasted on diesels would certainly have been better spent renovating the atrocious track over which both steam and diesels wracked their spring rigging, and the end result is predictably a much reduced system—in fact by the year 2000 there will probably be no 3ft 6in gauge remaining in Western Australia. A little will be converted to 4ft 8½in, but most will be torn up, for to American diesel and oil interests, 1,000 tons of freight is more profitably (to them) handled by one-hundred ten-ton road trucks than by a single railway train, even if the latter is diesel.

WESTERN AUSTRALIA: PRINCIPAL DIMENSIONS OF SELECTED LOCOMOTIVES

Class	*Type*	*Cylinders in× in*	*Wheels ft in*	*Heating Surface sq ft*	*Super-heating Surface sq ft*	*Grate Area sq ft*	*Weight tons*	*Pressure lb/sq in*	*Tractive Effort at 85% (lb)*
A	2-6-0	12 × 20	3 3	541	—	9·8	19·6	140	8,790
B	4-6-0T	14 × 21	3 1¾	595	—	11·6	35·3	160	15,140
C	0-6-0ST+T	10½ × 18	3 0	713	—	12·4	19·0	120	6,150
CA	4-6-2	16½ × 22	4 1	1250	—	17·8	42·1	175	18,200
DS	4-6-4T	18 × 23	4 6	971	174	18·6	71·2	160	18,770
DD	4-6-4T	18 × 23	4 6	1145	174	18·6	72·0	160	18,770
E	2-4-4-2T	(4) 10 × 18	3 3	836	—	13·0	33·5	120	9,420
E	4-6-2	17 × 23	4 6	1420	—	19·0	48·6	180	18,330
ES	4-6-2	19 × 23	4 6	1097	183	18·8	48·7	160	20,911
F	2-4-2T	10 × 18	3 3	418	—	6·5	23·0	120	4,710
F	4-8-0	17 × 23	3 6½	1361	—	18·8	54·1	175	23,260
FS	4-8-0	17 × 23	3 6½	1097	183	18·8	55·5	175	23,260
G	2-6-0	14½ × 20	3 3	781	—	13·7	25·2	160	14,650
G	4-6-0	14½ × 20	3 3	781	—	13·7	26·8	160	14,650
J	4-6-0	15 × 22	3 6½	686	—	11·5	28·5	140	13,850
K	2-8-4T	17 × 21	3 2	973	—	15·9	53·0	160	21,750
L	4-6-2	18 × 22	4 6	1066	214	20·5	51·0	160	17,950
M	2-6-0+ 0-6-2	(4) 12½ × 20	3 3	1340	—	22·6	68·8	175	23,850
MSA	2-6-0+ 0-6-2	(4) 13¼ × 20	3 3	1086	180	27·0	74·0	160	24,488
N	4-4-4T	15½ × 21	4 1½	829	—	15·8	48·8	160	13,880
O	2-8-0TT	15½ × 21	3 1½	862	—	15·8	36·4	160	18,320
P	4-4-0	15 × 20	4 6	869	—	13·9	28·0	135	9,550
*PR	4-6-2	19 × 26	4 6	1494	354	35·0	63·3	175	25,855
Q	4-6-0	14½ × 20	3 3	875	—	17·0	34·0	160	14,950
Q	4-6-2T	15 × 22	3 6½	817	—	14·0	39·0	160	15,800
R	4-4-0	16 × 22	4 9	901	—	16·6	31·8	160	13,450
RA	4-4-2	16 × 22	4 9	901	—	16·6	33·1	160	13,450
S	4-8-2	19 × 24	4 0	1673	448	40·0	75·3	200	30,685
T	4-4-0	15 × 20	4 4	846	—	15·2	29·8	140	10,300
U	4-6-2	18 × 24	4 6	1125	260	26·0	58·0	180	22,032
UT	4-6-4T	18 × 24	4 6	1125	260	26·0	79·8	180	22,032
V	2-8-2	19 × 26	4 3	1817	492	40·0	80·7	215	33,630
W	4-8-2	16 × 24	4 0	1117	305	27·0	63·3	200	21,760

*Also PM, PMR.

Chapter 8

TASMANIA

Tasmania is Australia's only island state, a beautiful isle set to the south of Victoria from which it is separated by the Bass Strait, flowing with ordinary sea water and not, as is sometimes fondly imagined, with best Burton bitter beer! Although only a small island state, Tasmania thought big in the manner of early railways, and the Launceston & Western Railway was built to the 5ft 3in gauge. This broad gauge line had only five locomotives during its whole existence, three 4-4-0 and a 4-4-0T, all by Stephensons, with inside cylinders and outside-framed bogies. They were essentially the same as the H and D classes in South Australia, differing mainly in dimensions. The odd engine was a Sharp Stewart 2-4-0 with outside cylinders, probably built for another customer and rebuilt in 1890 as a 3ft 6in gauge 4-2-2 as A class No 1 by the TGR. All these were built in 1872, and had to suffice for all traffic on the L & W (which ran from Launceston to Deloraine) until broad gauge was abolished in 1888. The four Stephensons were put up for sale; two went to Victoria to contractors and one later to SAR as a tender engine.

The first narrow gauge line on the island was the Tasmanian Main Line Railway, from Launceston to Hobart, opened finally in 1876, after about five years of construction work. The TMLR effectively fathered all remaining lines on the island, being of the useful 3ft 6in gauge adopted later for most routes in Tasmania. The TMLR had a diverse collection of locomotives, first of which were five Hunslet 4-6-0Ts built in 1874, followed the next year by two 4-4-0T with larger wheels, for passenger work. All these Hunslet locomotives were distinguished by inclined outside cylinders, inside valves and motion, sloping-front smokeboxes curved out laterally to embrace the cylinders, and disc bogie wheels. The first tender engines comprised a trio of outside-framed 0-4-2s by Neilson, with horizontal outside cylinders and four-wheeled tenders. Not surprisingly, they did not ride well, and were converted to 4-4-0s. Three new 4-4-0s with inside frames were purchased from Hunslet in 1884, sturdy little engines lasting until 1929. The final engines ordered for the main line company before its absorption by the later Government railway, came from Dübs, and comprised two 4-4-2Ts for local work and four sturdy 4-6-0 tender engines, the two classes coming out in 1886 and 1889 respectively.

The Tasmanian Government Railways came later on the scene, and the first of their 3ft 6in gauge motive power stock comprised fifteen delightful Beyer Peacock 4-4-0, later classed B, and twenty-eight C class 2-6-0, of the standard Australian type, each class being introduced in 1885. The real A class, as distinct from No 1, the rebuilt broad gauge 2-4-0 as a 3ft 6in 4-2-2, consisted of eight Beyer Peacock 4-4-0s Nos 2–9 built in 1892 as an enlargement of the B class, and these fine locomotives worked main line expresses until finally ousted by R class Pacifics in 1923, although the As survived on secondary duties until after the second world war, as indeed did the two earlier classes. For suburban work, Beyer Peacock continued their original monopoly by building five 2-4-2Ts, also in their classical mould, and very much an extended 'Isle of Man' 2-4-0T. These were allocated class D, and lasted into the 1950s.

Across the turn of the century, when most railways in the world were frantically building bigger and better power, introducing superheaters, compounding, piston valves, and other developments, the TGR experienced a hiatus, and fifteen years were to elapse before the next new class of locomotive appeared, in 1907. This was the E class 4-6-0 from Beyer Peacock, distinguished by horizontal cylinders, overhead piston valves actuated by Walschaerts gear, and boilers with Belpaire fireboxes. Only two were built, but soon afterwards Belpaire boilers began to be applied to the A and C classes, these rebuilds being distinguished also by pitching the boilers higher and fitting extended smokeboxes. As thus modified, seven of the 4-4-0s became class Ab, and six 2-6-0s class CC.

Another five years elapsed before further 3ft 6in engines were acquired, although in the interim engines had been supplied for the 2ft 0in

In February 1971 the Tasmanian rail centenary was celebrated by running thirty-one steam trains. Veteran 2-6-0 CCS23 pilots a Class Ma 4-6-2 out of Launceston towards Ross. *(J. Joyce)*

gauge North Dundas Tramway on the mountainous west coast. These will be covered later, but the last of these were the world's first Garratts, the K class of 1909, and their success led to the purchase of four Garratts for the main 3ft 6in gauge system. Accordingly there arrived from Gorton two 2-6-2+2-6-2, class L, for freight work, and the two remarkable eight-cylinder 4-4-2+2-4-4 express engines, which latter astounded all by travelling at 55mph, a speed at which they rode with incredible smoothness. One of the double Atlantic Garratts was involved in the Campania smash in 1916, when the Launceston to Hobart express derailed on a horseshoe bend, killing the driver and seven passengers. One opinion by a mining engineer blamed the disaster on the poor state of the track, but the official reason, uncontested by the poor defunct driver, was 'excessive speed', this being blamed on the Garratt's smooth riding properties, claimed to disguise the actual speeds attained. As a result, the Garratts were put in disgrace and relegated to freight service, expresses being again entrusted to the rebuilt A 4-4-0s until such time as Pacifics were available.

The M class Garratts were taken out of service about 1930, but lay derelict at Launceston until 1950–51, when finally scrapped. The L class lasted until 1937, being of simpler design and in any case largely confined to slow freight trains, and after a period out of use were reinstated from 1943 to 1946 to help with wartime and post-war traffic. Highly successful Garratts were later introduced on the private Emu Bay Railway, but except for a batch of ASGs, the TGR was not disposed to continue with their pioneer work on the Garratt locomotive type.

To cope with an urgent motive power shortage in the early 1920s six South Australian T class 4-8-0 were purchased and placed in service in 1921, plus an odd 2-6-2T from a timber tramway in New South Wales. Whilst these were at work,

Fitted with second-hand Garratt wheels for freight work, Class Ma No 4 hauls an AHRS tour on the Herrick line in 1970.

(*D. Allerton*)

Above: Beyer Peacock Class E 4-6-0 at Devonport in 1934.

(*J. Buckland*)

two new notable classes were in course of design and construction, and 1922 saw the first new Q class 4-8-2 arrive from Perry Engineering of Adelaide. These splendid machines were the first really modern power on any Australian narrow gauge line, and the first 4-8-2 type in Australia. A large boiler with wide Belpaire firebox supplied superheated steam to two outside cylinders through overhead inside-admission piston valves, actuated by Walschaerts gear. A high running plate added to the modern appearance, and so successful were they that nineteen were built by 1937, the last batch from Clyde with higher pressure and tractive effort such that their haulage capacity almost equalled the L class Garratts, which they replaced. Contemporary with the Q class and delivered in 1923, came four Pacifics of class R, of similar features and with many components interchangeable with the 4-8-2s. Again, not quite so powerful as the M class Garratts, they were sufficient to handle all main passenger trains; with two cylinders against the Garratt's eight, no articulated joints, and no inside motion, the R class was clearly more economical in maintenance. Apart from being the first Pacifics in Tasmania, the R class locomotives were also the first wide-firebox 4-6-2s in the whole of Australia, a fact which perhaps few appreciate. Tasmania at that time was thus in the forefront of Australian motive power practice, although the lead was not maintained, the other states soon catching up and overtaking the island system.

The sound and simple Q and R classes served Tasmania's needs for many years and other than the Garratts were the largest steam locomotives ever used on the TGR. After their introduction, it was nearly thirty years before any new designs were built specifically for Tasmania, but types new to the system appeared sporadically, from various causes. In 1924 the CCS class 2-6-0 came out, third in line from the original Beyer Peacock Mogul. While the CC class was the original Gorton product upgraded with high-pitched Belpaire boiler, the CCS added superheater, outside Walschaerts valve gear, and new cylinders with piston valves. In 1939, eight Wf class 2-6-4Ts were purchased from New Zealand and these replaced the old D class 2-4-2Ts on suburban services. Their new owners classed them DS. Then in 1944, when wartime traffic had strained the system's resources to the utmost and caused the resuscitation of the L class, some fourteen ASG 4-8-2+2-8-4 Garratts were purchased from the CLTB, the last going into service in 1946. On an average, the ASGs remained about ten years on the TGR, the last being taken out of service in 1957. Two were later sold to the Emu Bay

Overleaf: Class M 4-6-2 No 3 passing Avoca with a train for St. Mary's, in the centennial week.

(*D. Allerton*)

M
3

Railway where they proved unpopular, being in original ASG design condition, whereas the EBR had substantially rebuilt and improved its earlier ASGs.

The final steam classes built for Tasmania came out in 1951, and comprised ten light 4-6-2s, class M, by Robert Stephenson & Hawthorn (based on the Indian YB class, substantially smaller than the previous Pacifics), plus eight 4-8-2 from Vulcan Foundry, slightly smaller than the Q class, and based on similar engines supplied to the Gold Coast (later Ghana) Railways in West Africa. Thoroughly modern engines, the M and H classes differed in features due to their diverse ancestries, the 4-6-2s having plate frames and Belpaire fireboxes, while the 4-8-2s used bar frames and round-topped fireboxes. Both designs had SCOA-P wheel centres throughout, and roller bearings on all axleboxes and crankpins. Undoubtedly the most able of Tasmania's steam locomotives, they had the misfortune to arrive simultaneously with the first diesels, and were almost immediately relegated to secondary duties, such as pick-up freight turns, making small utilisation of their capacity for extended service with little attention. With their

A pair of Class H 4-8-2s near Bishops Bourne, en route for Devonport, in 1971. (*D. Allerton*)

extensive application of roller bearings, these final Tasmanian classes were amongst the most modern in concept through the whole of Australia; in this particular respect they were *the* most modern, as even the NSW 60 class Garratts had rollers only on the main driving crankpins. With virtually no passenger work to perform, four M class were in 1957 given smaller wheels taken from scrapped ASG Garratts, and classed MA, this alteration giving them a little more tractive effort.

In 1971 the Tasmanian Government Railways celebrated their centenary, and after being almost fully dieselised for fifteen years, astounded all by laying on a programme of steam-powered passenger trains during February, the centennial month, using eight serviceable steam locomotives of classes H, M, MA, and the venerable CCS No 23. Most engines were freshly painted in a livery of post office red, one or two remaining green. For much of the period Tasmania's notoriously wet weather gave way, leaving the sun to shine brightly on red engines and green countryside alike. It was a fitting end to Tasmanian steam power, ending with a bang rather than the usual whimper, and delighted hundreds of steam-starved Australian railfans, who swarmed across the Bass Strait to ride, record,

relish and photograph more steam passenger trains than had run on the mainland for years. At the time of writing, a few steam locomotives are just serviceable, and able to perform light duties on reduced boiler pressure, but to all intents and purposes, steam finished in Tasmania exactly one hundred years after it started.

As well as the broad and 3ft 6in gauge engines, the TGR also owned and operated ten 2ft 0in gauge machines on the North East Dundas Tramway. Four were typical Krauss 0-4-0 well tanks of the type seen all over the world, while three larger 0-4-2T came from Sharp Stewart, in 1897. The Krauss type, class H, were second-hand, and had outside Stephenson gear, disc wheels, and inside frames integral with the water tanks, while the Scottish engines had outside frames and Walschaerts valve gear. These little engines were hard pressed to cope with mineral traffic over heavy grades, and in 1901 Hagans of Erfurt, Prussia, built one of their patent semi-articulated engines for the line. These were difficult to classify by White's notation, having a leading pony truck followed by a pair of cylinders driving six wheels in the normal manner. Behind came four more wheels able to swing in a pivoted frame, but coupled to and driven from the rigid group by a pantograph motion. For traction purposes they were 2-10-0T, all driving wheels being coupled together, but as a vehicle the engine was a 2-6-4-0T. The Hagans engine was class J and a fitter's nightmare, hence it was no surprise that it was decided to use something else where more power was needed. The final 2ft 0in engines acquired were the well-known K class, the world's first Garratt engines, described and illustrated in *The Garratt Locomotive*. The K class survived the closure of the line and K1, the very first Garratt, was returned to Manchester, and upon closure of Beyer Peacock's works transferred to the Festiniog Railway in North Wales. Here it is too big to operate through the Festiniog's cramped loading gauge, and persistent rumours have abounded that the engine is to be cut down to suit. Such mutilation of a historic locomotive would be nothing short of vandalism, and it is to be hoped that some other way of steaming K1 will be found, even if this means loaning the locomotive to some other preservation group of the right gauge.

Tasmania had a high proportion of non-state lines, and these will be found in chapter 9.

TASMANIA

Class	*Type*	*Cylinders in × in*	*Wheels ft in*	*Heating Surface sq ft*	*Superheating Surface sq ft*	*Grate Area sq ft*	*Weight tons*	*Pressure lb/sq in*	*Tractive Effort at 85% (lb)*
A	4-4-0	15½ × 22	4 7		—			150	12,250
B	4-4-0	14½ × 20	4 0		—			140	10,400
C	2-6-0	14½ × 20	3 3		—			140	12,800
CC	2-6-0	15 × 20	3 3		—			175	17,150
CCS	2-6-0	15 × 20	3 3		—			175	17,150
D	2-4-2T	12½ × 20	3 9		—			140	8,250
E	4-6-0	16 × 22	3 6		—			175	19,900
F	2-6-0	14½ × 20	3 3		—			185	16,950
L	2-6-2+ 2-6-2	(4) 15 × 22	3 6	1686	333	33·9	90·0	160	30,800
M	4-4-2+ 2-4-4	(8) 12 × 20	5 0	1686	333	33·9	94·6	160	26,100
P	2-6-2T	14 × 20	3 0		—			175	16,200
Q	4-8-2	20 × 24	4 0		—			180	30,500
R	4-6-2	20 × 24	4 7	1534	382	32·5		160	23,700
M	4-6-2	16 × 24	4 7	1109	265	23·1	54·2	180	17,100
MA	4-6-2	16 × 24	4 0	1109	265	23·1	54·0	180	19,500
H	4-8-2	18 × 24	4 0	1588	372	34·0	66·4	200	27,500
*G	0-4-2T	12 × 16	2 6		—			140	9,100
*H	0-4-0T	7 × 12	2 0		—			175	3,650
*J	2-6-4-0T	15¾ × 15¾	2 7½		—			195	20,500
*K	0-4-0+ 0-4-0	(2)11 × 16 (2)17 × 16	2 7½	628	—	14·8	33·5	195	16,300

*2ft 0in guage

Chapter 9

PRIVATE RAILWAYS

In Australia there were, and to some extent still are, many railways, excluding purely industrial works railways, not owned by the main state systems. These varied extensively in extent, character, operation and motive power, and in a work such as this only those with the most interesting locomotives may be dealt with more than fleetingly, this being a locomotive and not a general railway book. It is most convenient to cover these lines on a state-by-state basis.

New South Wales

This state had three private railways of particular interest, two mainly for the coal trade, and the third for iron ore. First and foremost is the South Maitland Railway serving the coalfields south-west of Maitland, itself a few miles inland from Newcastle. The SMR started life about 1890 as the East Greta Colliery Railway, gradually extending and connecting other colliery lines until a substantial network of one main line and several branches was in operation. The South Maitland Railways were formed in 1918 to take over the whole of this operation, which was far more than a mere colliery railway. From East Greta Junction on the outskirts of Maitland a

The major steam user today in Australia is the South Maitland Railway, near Newcastle, NSW, a coal-hauling line. Last built of the Beyer Peacock 2-8-2Ts, No 31, hauls the daily general freight near Weston. (*A. E. Durrant*)

fully-signalled double-track main line extended to the town of Cessnock, and apart from coal there was general freight and passenger traffic, including at one time a through passenger train from Cessnock to Sydney. In this respect, the SMR and its precedessors were rather similar to the smaller railways in the *old* South Wales, perhaps the Cardiff Railway, or the Rhondda & Swansea Bay Railway.

Over the years, more than thirty locomotives graced the line's roster, comprising some standard industrial tank engines, some former main line locomotives, and other more interesting machines designed and built expressly for the railway. The first engine was an ex-main line 4-4-0T, followed in 1900 and 1902 by two standard Avonside saddle tanks, one 0-4-0ST and one 0-6-0ST. These were the last standard industrial engines bought by the line, and all subsequent locomotives had some feature of interest.

There were three Kitson 0-6-0 tender engines dating from 1877–79, acquired second-hand, which were virtually identical with the Taff Vale Railway K and L classes, used for coal haulage in the old South Wales. As an extension of their usual four- and six-wheeled engines, Avonside built two 0-8-0ST in 1903–04, with long connecting rods driving the third coupled axle. The same firm in 1908–11 supplied three 0-8-2T with Walschaerts valve gear, which were a great leap forward from the usual British industrial tank either of that period, or for that matter, until the end of industrial steam.

For passenger work, the line bought two outside-cylinder 4-4-2Ts from the NSWGR, their class 13, and at a later date a further engine of the same wheel arrangement, with inside cylinders, class 11. The principal passenger engines were three 4-6-4Ts to the government railway's S (30) class design, but built new by Beyer Peacock in 1912 and 1923.

The principal type of engine on the line was the Beyer Peacock 2-8-2T, designed as a tank version of the NSWGR T (50) class 2-8-0, a Lucy design also first built by Beyer Peacock. These were far and away the best coal haulers on the line, and fourteen were built between 1912 and 1926. Eventually the older engines were disposed of, leaving the SMR with three 4-6-4Ts and the fourteen 2-8-2Ts, a well standardised stock for such a railway. In 1961 three diesel railcars were bought for passenger work, but not surprisingly,

0-6-4T No 5 of J & A Brown's coal-hauling railway in New South Wales was one of Beyer Peacock's original condensing locomotives of the Mersey Railway in England. (*A. E. Durrant*)

Brown's 2-8-2T No 10 *Richmond Main* sizzles in the yard at Hexham in 1972. Very similar to Robinson's Great Central 2-8-0, it in fact preceded that design. (*A. E. Durrant*)

this heavy investment was more than the traffic could bear and services ceased in 1967, leaving the railway once more 100 per cent steam-operated, but now only a coal hauler, with a daily general freight train to Cessnock. In early 1978 this condition still prevailed, and the SMR bade fair to be the last outpost of regular steam operation in the whole of Australia.

J. & A. Brown—Richmond Vale Railway

Competing in the same coalfield as the SMR was the Richmond Vale line, owned and operated by the Brown organisation mainly as a service to the company's coal mines.

Originally there was only a colliery at Minmi, with a short connecting line across the swamp to Hexham where the Hunter River Railway was joined, and for this two 0-4-2Ts with inside cylinders were acquired. Built by Hawthorns of Newcastle in 1856, they were reputedly for war service in the Crimea, and available for purchase just when John Brown needed them. A larger Kitson 0-6-0ST was supplied new in 1878, and another of the same type built 1870, bought second-hand from the NSWGR in 1891. So far, the motive power had been of no great interest, and the turn of the century saw Brown's using three engines on the Hexham to Minmi railway plus the fourth shunting at Pelaw Main colliery, where traffic was fed into the East Greta Railway. Apparently this was not to Brown's liking, and the Richmond Vale Railway was built through scenic terrain to connect Minmi with Pelaw Main, enabling Brown to handle all his coal traffic to Hexham interchange, using his own locomotives.

More motive power was needed for this extra traffic, and after a short period of hiring government engines, Brown bought four of the former Mersey Railway 0-6-4T, rendered redundant by the electrification of that line. Here again we have a parallel with the old South Wales, for most of the other Mersey tanks, seven 2-6-2Ts and three 0-6-4Ts, were bought by the Alexandra (Newport & South Wales) Docks & Railway Company about the same time, and used also for coal haulage from Pontypridd to Newport. Brown's ex-Mersey locomotives were placed in service between 1907 and 1909, but meanwhile a much more modern 2-8-2T had been ordered from Kitson's. Built in 1908, this was a thorough-going Great Central 2-8-2T, except that the GCR never owned such an engine! A likely explanation seems to be that Kitson built a large number of Robinson's 0-8-0 freight engines for the GCR, in the years 1904–07, and would thus have available all the necessary drawings and patterns for their construction. Faced with a request to build a powerful tank engine for Brown, it was a quick drawing office job to take the 0-8-0 as a basis, add leading and trailing pony trucks, side tanks and bunker, in fact, an instant 2-8-2T. This conjecture seems confirmed by the basics of the locomotive being Great Central, while the tanks, bunker and cab were of pure Kitson design. Two more of these engines were

Brown's principal main line locomotives were former ROD 2-8-0s of Great Central Railway design, acquired after the first world war. No 23 blasts out of Stockrington with wood chaldrons of coal, in 1971.
(*A. E. Durrant*)

built in 1911, and were the last and largest *new* locomotives built for Brown. After the first world war a substantial upsurge in traffic necessitated further motive power, and thirteen redundant 2-8-0 tender engines were bought from the British Government, these being former ROD engines built to Robinson's Great Central class 8K, better known as LNER class O4. It is strange that these GCR engines should become available to a company whose previous locomotives should also by chance contain so many GC components, especially as the previous (0-6-4T) machines had come from the same part of England. Equally strange to relate, there were ten NSWGR design 2-8-0s available from the ROD at the same time, but whereas the Australian type engines went to Belgium, English type 2-8-0s came to Australia.

Other locomotives owned by Brown's railway were odd tanks acquired second-hand from time to time, and as at early 1973 there were only two RODs barely serviceable, two of the Kitson 2-8-2Ts, and a former Government 4-6-4T No 3013, third-hand from Hebburn No 3 colliery. However, this situation was not difficult, for only one on-line colliery remained, that at Stockrington, only a couple of miles past Minmi, and most of the railway lay derelict with track torn up. Nevertheless, there will be work for the Hexham engines for some time yet, if only shunting wagons to and from the exchange sidings to Brown's washing plant, many of which wagons arrive from other mines and proceed to Newcastle seeing no more of Brown's line than the sidings and plant.

Hebburn Collieries

This was one of the most important coal mines in New South Wales, and earns a place in this book by possession of one unusually interesting locomotive. So far, two interesting connections between Lancashire and the Newcastle coalfields have been covered—the former Mersey Railway 0-6-4Ts bought by John Brown, followed by the Gorton-designed 2-8-0s for the same concern. Hebburn's locomotives tended to be uninteresting with one exception, which added a second link, perhaps the most unlikely of all, between the Mersey Railway and the Maitland coalfield. The original Mersey Railway had two main locomotive classes both by Beyer Peacock, the 1885 0-6-4T with inside cylinders, probably unsatisfactory when running forward, and the later 1887 outside-cylinder 2-6-2T whose pony trucks both ends ensured good tracking in each direction. Apart from the 0-6-4Ts sold to Brown, ten engines including both types were bought by the Alexandra Docks & Railway Company in South Wales, and the 2-6-2Ts were sufficiently useful for two more of the same general dimensions, but modernised in detail, to be ordered from Hawthorn Leslie. These were delivered in 1920, the last new steam for the line, and almost immediately assumed new identities as Great Western Railway Nos 1205–06. They survived into British Railways days, one being scrapped in 1951, the other in 1956. Just as the second was withdrawn, Robert Stephenson & Hawthorn (successors to Hawthorn Leslie) were building an identical engine for Hebburn, NSW! This engine had less fortune than its predecessors, for after cracking a cylinder it was found cheaper to buy a 30 class 4-6-4T from the NSWGR and the 2-6-2T was then scrapped.

The Silverton Tramway (*3ft 6in gauge*)

The Silverton Tramway came about directly as a result of Australia's interstate parochiality. The South Australian narrow gauge line from Port Pirie extended inland to Cockburn on the New South Wales border, and was heading for Broken Hill where ore deposits provided an ideal source of heavy traffic. However, New South Wales was not going to allow the 'Crow Eaters' to run a railway in *their* state, so finally a company was formed to build and operate the necessary narrow gauge extension. For some legal reason it had to be called a 'tramway', although in fact it was a proper railway, and eventually formed a link in the main transcontinental route between Sydney and Perth, possibly the only 'tramway' in the world to achieve such eminence.

Most of the locomotives on the Silverton Tramway were standard Beyer Peacock 2-6-0s of the South Australian Y class, which they met on the border at Cockburn. For shunting and local work there were two of a 2-6-2T version, these being another Beyer Peacock standard, familiar to European enthusiasts on the Alcoy Gandia line in Spain. The only type designed for the Silverton Tramway comprised four Beyer Peacock 4-6-0s, class A, extremely handsome engines with Belpaire fireboxes and outside Walschaerts gear driving slide valves. Two were supplied in 1912 and another pair in 1915, and they were the company's principal main line power for nearly forty years. The company's final steam engines, also from Gorton in 1951, were four 4-8-2s to the same design as Western Australia's W class,

The Silverton Tramway, connecting Broken Hill with SAR's narrow gauge, had four delightful Beyer Peacock 4-6-0s, one of which is preserved at Mile End, SA.

(A. E. Durrant)

under construction by Beyer Peacock at that time. The original WAGR W class had 'skyline' casings extending over the boiler top, this being the hallmark of F. Mills, the WAGR Chief Mechanical Engineer. However, before the engines were completed Mills died, and his successor, C. W. Clarke, discarded the skyline casing, much to the improvement to the locomotives' appearance. Meanwhile, the Silverton Tramway had ordered four Ws, and for some reason they were delivered to the original design which never saw the light of day in Western Australia. The Silverton W class had a short life, the line being dieselised in 1960–61, and it was finally replaced altogether by the new standard gauge link from Broken Hill to Port Augusta. The name, Silverton, incidentally, comes from the small town half way along the line, and was presumably chosen to disguise the real *raison d'être* of the line. Although only numbering twenty-five in total, three of the company's engines are preserved in museums, a W and an A at Mile End, and another W at Menzies Creek.

Victoria

Most private lines in Victoria were purely industrial concerns, plus a few agricultural tramways which have no place in this book. Most interesting from a locomotive point of view was the Australian Portland Cement Company's line at Fraynsford, which used Garratts, firstly of the WAGR M, and later of the ASG type.

South Australia

Again thin on private lines, South Australia did have one notable line, the Whyalla & Iron Knob Tramway, a 3ft 6in gauge ore carrier owned and worked to supply Whyalla steelworks and also to its east coast steel plants at Newcastle and Port Kembla. Formerly isolated on the Eyre Peninsula at the top end of Spencer Gulf, opposite Port Pirie, Whyalla was in 1972 connected to the rest of Australia's rail system by a standard gauge link to Port Augusta. However, this has not affected the W & IK, with its thirty-one-mile main line, plus a sixteen-mile branch to Iron Baron. The most notable locomotives owned by the line were some Baldwin 2-8-2s of considerable size, somewhat larger than the WAGR V class.

Queensland

The 'Banana state' had no privately-owned railways with locomotives of particular interest other than a few Shire Tramways, operated by the local shires to provide services where the QGR thought railways not worth building. Some of these were 3ft 6in gauge, and used ex-QGR engines, but one 2ft 0in line, the Port Douglas Shire Tramway, had an Orenstein & Koppel 0-4-4-0T Mallet, similar to those in Tasmania. Last and largest, the sugar cane fields of northern Queensland operated extensive tramways to the mills, many of these in use today, some still with steam power. The market for small engines was at one time so great that the Leeds firm of John Fowler set up a subsidiary at Bundaberg.

Western Australia

Western Australia had a large number of private lines, of which two were of a main line nature. Of these, the Great Southern to Albany, and the Midland Railway of Western Australia were each absorbed by the WAGR. The GSR line was taken over at an early date, and its locomotives covered in chapter 7, but as the MRWA was not taken over until 1964, by which time it was fully dieselised, its steam story belongs to the private railways chapter. The MRWA extended 277 miles, from Midland Junction, ten miles from Perth, to Walkaway, twenty-nine miles south of Geraldton, and like the Great Southern, was built on the land grant principle.

The first MRWA locomotive, class A, was a strange outside-framed 2-6-2 reputedly named *Walkaway*, built by R. W. Hawthorn in 1887. The engine is described and illustrated in *The Engineer* for 10 June 1887, although as built it seems to have been named *Emigrant*. For its day it was quite a remarkable locomotive, with a large wide firebox boiler, and trailing axle well spaced, leaving room for a capacious ashpan. Outside cylinders and inside valves and gear were applied, but the weak point seems to have been the leading and trailing truck arrangement, with outside radial axleboxes, probably very stiff on curves. The engine did not last long in MRWA service, and was sold to a timber line in 1900.

The B class, although from the same builders, comprised 4-4-0s of the Beyer Peacock pattern, nine being built in 1891–92, followed by two more second-hand from the WAGR in 1912, these latter having been built by Martins in 1896. This was almost an Australian standard, these particular engines having been P class on the WAGR, and identical with the South Australian Z class. One of the English batch, painted in Midland red, is preserved at Geraldton, the only surviving member of MRWA's steam roster.

The C class locomotives, built by Kitson in 1912, were American-looking 4-6-2s with bar frames, with slightly inboard piston valves driven by Walschaerts motion. At the rear end a Belpaire firebox was supported in a cradle casting similar to that used in South Africa, and indeed the MRWA engines were probably based in this respect on the Cape Government 'Karoo' (5A) class supplied by the same builders in 1903. The last built of the C class, No 18, had a smokebox superheater which was later removed, but it is not clear whether the engines later had more modern superheaters fitted, most of the line's steam records having disappeared during dieselisation and subsequent absorption into the WAGR.

Built by Kitson, the Midland Railway of Western Australia Pacifics showed American and South African design features.

(*WAGR*)

Top: In contrast to the Pacifics, the MRWA 4-8-0s were of decidedly British design, but were Baldwin-built. (*WAGR*)
Above: MRWA's final steam class boasted cylinders of modern design and, like No 21, would have looked at home on any British colonial railway. (*WAGR*)

The D class on the Midland was the antithesis of the Pacifics, being American-built 4-8-0s of thoroughly British design, with plate frames and inclined cylinders, such that the basic design appears to be almost identical with the Nasmyth Wilson engines built for the Lagos Government Railway in 1903. Exactly why Baldwin had to build two African-type engines, of English design, for Australia has never been explained by the local historians.

The new A class, comprising nine 2-8-2s by Kitson, appearing in threes alternate years from 1925 to 1929, were the Midland's last new steam power. These were straightforward, modern light Mikados with large diameter long-lap valves, making them among the most modern engines in Australia when built, certainly more modern than anything on the neighbouring WAGR. Steam traction effectively ended on the Midland in 1958, although a number of engines were retained in reserve until 1963, the year before absorption by the Government.

The Jarrah Belt Railways

In Western Australia's southwest corner there are extensive forests of such trees as jarrah, karri, and tuart, jarrah being renowned worldwide as the best timber for railway sleepers. To exploit these areas Millars, a large timber firm, built extensive railways into the forests, these all being to 3ft 6in gauge, lightly-laid, and connecting the forestry workings with the company's sawmills and the WAGR. In 1930 there were eight major rail systems owned by Millars, and these comprised:

Yarloop	63 miles
Mornington	44 ,,
Jarrahdale and Rockingham	57 ,,
Wellington	17 ,,
Marrinup	4 ,,
Kirup	26 ,,
Jarrahwood	13 ,,
Canning	15 ,,

This extensive system owned twenty-five locomotives at the time, and there were probably as many again owned by various other concerns such as Bunning Brothers, in the same area. Many were second-hand engines from the WAGR and other state concerns, the Beyer Peacock 2-6-0 being a favourite type. Unfortunately all of these fascinating lines have been closed down, replaced by road tractors. The last was the rump end of the Yarloop section, where Millar's G class 4-6-0 No 71 was the last steam locomotive in regular service in Western Australia.

Goldfield railways

There were a number of narrow gauge lines in the Kalgoorlie area, serving the goldfields there, and two locomotives of interest worked on them. The first was a Mallet 0-4-4-0T from Mount Magnet in Tasmania, one of two by Orenstein & Koppel, now preserved as the last of Australia's three Mallets in a park at Mussel Pools in the Swan Valley. The other goldfields area engine worth mention contained no mechanical features of interest, but was an 0-4-2T built for the private railway of 1ft 8in gauge, serving the 'Sons of Gwalior' mines, by the state railways workshops at Midland Junction.

Tasmania

This island paradise was equally a paradise for lovers of obscure private railways, for a full sixth of the mileage was in private hands. Most of this was of the standard 3ft 6in gauge, but there were also some 2ft 0in lines. In addition to the two narrow gauges, there was a rack line. It is not possible in this book to attempt coverage of all these delightful lines and their locomotives but the two principal lines will be dealt with.

The Emu Bay Railway

This fascinating title covered the 3ft 6in gauge line from Burnie on the Emu Bay, northern Tasmania, south through the mountains to Zeehan, centre for the rich mining area near the west coast. The line's power started with two Hunslet 4-4-0s dated 1883, followed by a Neilson 0-6-4T of 1888, which apart from outside Walschaerts gear, was very much a Drummond engine. The cold wet climate of 'Tassie' seemed home from home for Scots and their engines, as we will see in the following paragraphs. The two 4-4-0s were scrapped early on, but the 0-6-4T lasted until 1940, and was photographed in contrast with the first Garratts on their arrival. In 1897 there came the first two of the inevitable Beyer Peacock design 2-6-0, although these were by Martin of Gawler; there were eventually six of this type, all by Martins except for one genuine Gorton product. The first 'big' engines were three 4-8-0s by Dübs built 1900, followed in 1911 by a fourth from the North British Locomotive Co. which succeeded Dübs. With David Jones as consulting engineer, these had a thoroughly Highland Railway outline, and had the HR ever progressed from Jones' famous 4-6-0 to an eight-coupled design, they would have looked exactly like the Emu Bay engine. Dimensionally, they were very similar to South Australia's T class, which came later and may have been based on the Tasmanian engines. In overall features, inclined outside cylinders and inside valve gear, they were clearly derived from the Cape Government Seventh class, basic inspiration for so many 3ft 6in gauge 4-8-0s. After replacement by Garratts, these little 4-8-0s were out of use for many years until about 1960, when of all surprising things road-bound tourist traffic caused their resuscitation! How this came about concerns a 'round Tasmania' tour by coach, of which the central western section, from Guildford Junction to Rosebery, was devoid of roads, the Emu Bay Railway filling in with a delightful train comprising refurbished 4-8-0, complete with smoke deflectors, and three or four ancient carriages, all painted in two shades of blue. Upon completion of the Murchison Highway, tourists now go all the way by motor coach, thus missing a train journey which must have been one of the trip's highlights.

From the North Mount Lyell Railway, the EBR in 1928 acquired two Avonside 4-6-0s of 1899 vintage. According to *The Locomotive Magazine* for January 1903, which described and illustrated them, these too were designed by

Two of the Mount Lyell Railway rack locomotives are preserved. *Mount Lyell No 5* is at the 'Puffing Billy Museum' at Menzies Creek, Victoria, mounted on an inclined track with viewing passage underneath.
(*A. E. Durrant*)

David Jones, but they certainly did not have the Highland look, and were typical colonial 4-6-0s. These just preceded the line's first Beyer Garratt locomotives, three 4-8-2+2-8-4s which were the largest and most powerful locomotives to run in Tasmania. Based on the Kenya Uganda Railway EC1 class, these differed only in the water tanks, and other minor details, plus of course the gauge.

The final steam locomotives on the EBR were five Australian Standard Garratts, smaller than the Beyer Garratts, but which were extensively rebuilt and modified to eliminate, or at least reduce, the numerous faults which this class suffered. Of the five, only four were in use at one time, for the fifth, numbered 20A, was purchased from the TGR to replace No 20, destroyed in an accident. The Emu Bay Railway is now fully dieselised, the last steam operation being about ten years before these words were typed.

The Mount Lyell Railway

Operating also in Tasmania's ore mining area, the Mount Lyell Railway ran from Regatta Point, near Strahan on Macquarie Harbour, inland to Queenstown, and was also of 3ft 6in gauge. The main claim to fame was in this line being the first of only two rack-and-adhesion railways in Australia, and of its twenty-five route miles, four-and-a-half were on the Abt rack system. Locomotives for the adhesion section included Baldwin 0-6-0Ts and the 0-4-2T Abt engines were by Dübs. Although the Mount Lyell Railway is now but a memory, two of the rack engines have been preserved, No 1 at the West Coast Pioneer's Memorial Museum, at Zeehan, and the other (No 5) in the Puffing Billy museum, Menzies Creek, Victoria.

5501

Australia's only Giesl ejector application was to 'Pig' No 3616, the report of whose superb performances was banned by the diesel interests. Here in 1971, No 3616 in green livery heads an enthusiasts' trip through the Sydney suburbs.

(A. E. Durrant)

The final batch of Standard Goods 2-8-0 sported the ungainly Southern valve gear, seen clearly on No 5501.

(NSWGR)

BIBLIOGRAPHY

In preparing this book the author has consulted numerous publications containing relevant material, and for those desiring more detail than is possible to include in this work, readers are recommended to consult the books and periodicals listed below.

Australia

Railways of Australia. C. C. Singleton and D. Burke. Angus and Robertson. 1963.
Locomotives of Australia and New Zealand. J. F. Palmer.
Steam on the Downgrade. B. Andrews and R. Warren. 1967.
West by Steam. RTM. 1972.
Locomotives of Australia. Leon Oberg. A. H. and A. W. Reed Ltd. 1975.

New South Wales

Century Plus of Locomotives. ARHS (NSW). 1965.
New South Wales Steam Locomotive Data. NSWGR. 1970.
Flyer. RTM. 1970.
Tender into Tank. R. G. Preston. ARHS (Newcastle). 1970.

Victoria

Power Parade 1854–1954. VR. 1954.
North Williamstown Railway Museum. ARHS (Vic.).
The Era of the R Class Locomotives in the Victorian Railways.
The Era of the J Class Locomotives in the Victorian Railways.
The Era of the K Class Locomotives in the Victorian Railways.
The Era of the S and H Class Locomotives in the Victorian Railways.
(All written and published by Stephen Watson)

South Australia

Locomotives of the SAR. Marshall and Wilson. ARHS (SA). 1972.
500. D. Colquhoun, R. Stewien and A. Thomas. ARHS (SA). 1969.
600. D. Colquhoun, R. Stewien and A. Thomas. ARHS (SA). 1971.
700. D. Colquhoun, R. Stewien and A. Thomas. ARHS (SA). 1968.
Proceed to Peterborough. D. Colquhoun, R. Stewien and A. Thomas. ARHS (SA). 1970.
621. R. Stewien. ARHS (SA). 1972.
W. A. Webb. R. I. Jennings. Nesfield Press.

Commonwealth Railways

Locomotives of the Commonwealth Railways. R. Fluck, B. Marshall and J. Wilson. ARHS (SA). 1972.
Report of the Royal Commission appointed to inquire into Australian Standard Garratt. Government Printer, Perth, 1946.
Australia's Garratt. Robert Butrims. Geelong Preservation Society and ARHS (Vic).

Queensland

The Redbank Museum. J. Armstrong. ARHS (Q). 1970.
Lookout for Train (Pictorial). ARHS (Q). 1971.

Western Australia

WAGR Locomotives 1940–1968. A. Gunzburg. ARHS (WA). 1968.
One Hundred Years of Railways in Western Australia. G. J. Higham (ed.). ARHS (WA). 1971.
Steam Up. J. Joyce and A. Tilley. J. & A. Publications. 1972.

Tasmania

Railroading in Tasmania. T. C. T. Cooley.
Century of Steam (Pictorial). Zig Zag Press. 1971.

Private Railways

The Railways of the South Maitland Coalfield. G. H. Eardley. ARHS (NSW). 1969.
The Railways of J. & A. Brown. G. H. Eardley. ARHS (NSW). 1972.

Books with Australian items

The Garratt Locomotive. A. E. Durrant. David & Charles. 1969.
The Fairlie Locomotive. R. A. S. Abbott. David & Charles. 1970.
4-8-0 Tender Locomotives. D. R. Carling. David & Charles. 1971.
The Mallet Locomotive. A. E. Durrant. David & Charles. 1973.

Periodicals

ARHS *Bulletin.* Monthly, current.
Australian Railway Enthusiast. (ARE).
The Roundhouse. (RTM).
Light Railways.
The Locomotive Magazine.
The Engineer.
The Railway Gazette.

LIST OF ABBREVIATIONS

ARHS	Australian Railway and Historical Society.
ASG	Australian Standard Garratt.
CME	Chief Mechanical Engineer.
CLTB	Commonwealth Land Transport Board.
CR	Commonwealth Railways (now known as Australian National Railways, ANR).
CTC	Centralised Train Control.
EBR	Emu Bay Railway.
GA	Grate area.
GCR	Great Central Railway.
GWR	Great Western Railway.
HS	Heating surface.
Km	Kilometres.
lb	Pounds weight or force.
LMS	London Midland & Scottish Railway.
LNER	London & North Eastern Railway.
LNWR	London & North Western Railway.
LSWR	London & South Western Railway.
L & W	Launceston & Western Railway.
MRWA	Midland Railway of Western Australia.
NSW	New South Wales.
NSWGR	New South Wales Government Railways (now Public Transport Commission of NSW—Rail Division).
NZ	New Zealand.
lb/sq in	Pounds per square inch.
QGR	Queensland Government Railways.
ROD	Railway Operating Division (of Royal Engineers).
RTM	Rail Transport Museum.
SAR	South Australian Railways (now State Transport Authority of SA—Rail Division).
SHS	Superheating surface.
SMR	South Maitland Railway.
sq ft	Square feet.
ST	Silverton Tramway.
TAR	Trans Australian Railway.
TE	Tractive effort.
TGR	Tasmanian Government Railways.
TMLR	Tasmanian Main Line Railway.
UNRRA	United Nations Relief and Rehabilitation Administration.
USA	United States of America.
VR	Victorian Railways (now under the marketing name 'Vicrail').
WA	Western Australia.
WAGR	Western Australian Government Railways (now under the marketing name 'Westrails').
WIK	Whyalla and Iron Knob Railway.
WW I	First World War.
WW II	Second World War.

ACKNOWLEDGEMENTS

In preparing the manuscript for this book, the author has been most grateful to receive information and photographs, as requested, from all the major railway systems, who have proved most helpful, even though steam in some cases is long since defunct.

The various Australian enthusiast groups and their individual members have also been useful sources of information, to say nothing of inspiration and encouragement in a venture where the author's first thoughts were 'there's nothing new to say'. The author hopes that he has tackled the job in a manner which will provoke interest amongst established and well-informed Australian enthusiasts. as well as to fill a gap by providing basic information for readers overseas.

Finally, he thanks the Australian railway enthusiasts as a group, too numerous to mention individually, for the hospitality extended to his wife and himself, at society meetings, in their homes, beside the railway line; or in the most convenient pub.

INDEX